CONTENTS

Science can unveil the complexities of ourselves, and the world around us

THE CURIOUS COUNTRY

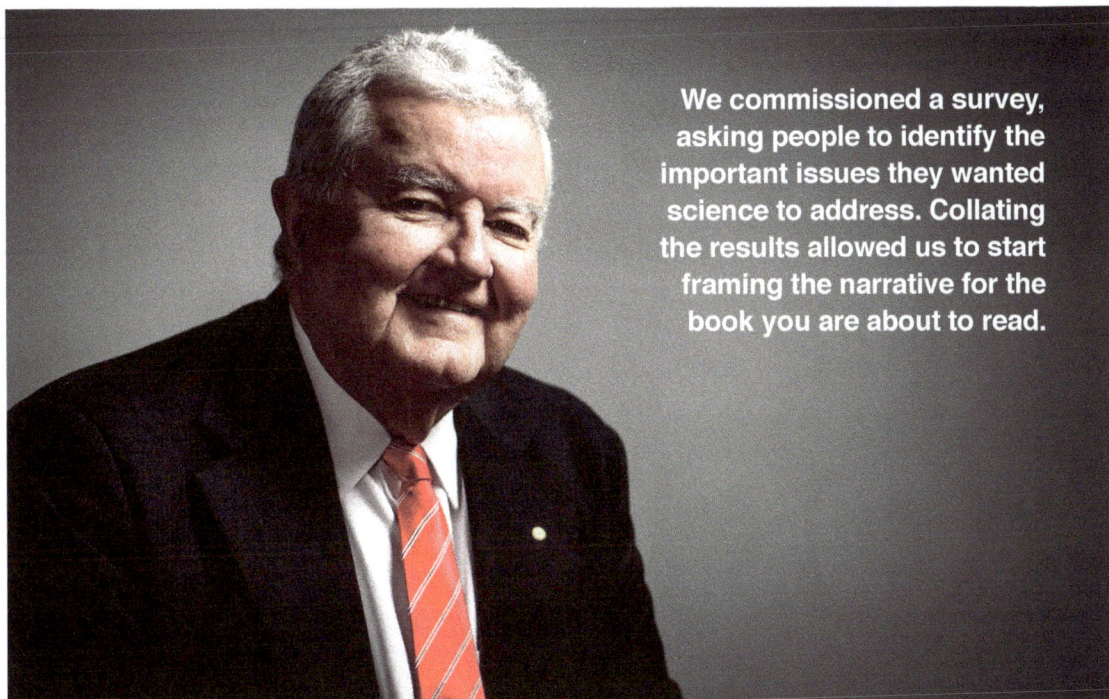

We commissioned a survey, asking people to identify the important issues they wanted science to address. Collating the results allowed us to start framing the narrative for the book you are about to read.

DURING 2012 I travelled the breadth and width of the country, made many speeches, gave many high-level briefings and attended many events and even more meetings. I met lots of people and was heartened by their interest in science.

It was immensely interesting, but I can't say it gave me a representative insight into how Australians view science and where their interests lie. Most of the people I met at these events had an interest in the specific topic we were discussing or had come to tell me whether they agreed or disagreed with some aspect of the science.

This led me to wonder what Australians more generally think about science. I know that the average Australian family enjoys the fruits of science every hour of every day; but does science intrigue them? Do they think about how science shapes the food they eat, the medicines they take and the technology they use? Are there issues in science that are not necessarily part of the topical agenda that intrigue them? Importantly, are there parts of science in the media rendered opaque by the tendency of scientists to talk in technical language?

What would Australians like to know more about? What, if anything concerns Australians about science? What inspires them?

So we asked them.

We commissioned a survey, asking people to identify the important issues they wanted science to address.

Collating the results allowed us to start framing the narrative for the book you are about to read.

We then set about finding the right people to write the book. They needed knowledge, creativity and passion. Leigh Dayton agreed to come aboard as editor. Both she and the authors have contributed hours and effort beyond what we or they imagined. For that I say thank you. I would also like to acknowledge the hard work of the staff in my office, without them, this book would not have seen the light of day.

Finally, to you, the reader, I also say thank you. As a citizen of the Curious Country, this is very much your book.

I hope you find it informative, but more than that, I hope you enjoy reading it. – *Ian Chubb*

SCIENCE MEETS SOCIETY

Scientists are an inquisitive lot, keen to tackle the scientific curiosities and concerns of Australians, writes Leigh Dayton

AS AUSTRALIANS recovered from the New Year's party whistles, fireworks and rounds of 'Auld Lang Syne', scientists were on the job at the Bureau of Meteorology in Melbourne. "It's [the record] going down," messaged one excited meteorologist as the nation's collective thermometer broke the all-time record. On 7 January 2013 the national average temperature for that day hit 40.33°C, beating the previous record of 40.17°C, set in 1972. Why? Why care? Who to ask?

The 'whys' of weather are complex. Who to ask about them is not. The answer is scientists. It's their job, their passion and their pleasure to tease out nature's secrets and apply their findings to the current events and queries of the day. From climate change to swine flu, astrophysics to zoology, this diverse group of men and women raise issues, seek answers and rigorously explore the curiosities and concerns facing today's world.

By definition scientists are an inquisitive lot. But what are the scientific curiosities and concerns on the minds of Australians? What worries them, baffles them, and sets their curiosity meter to 10 out of 10? To find out, the Office of the Chief Scientist (OCS) took the nation's intellectual temperature, surveying 1186 Australians: men and women aged 18 to 65, from all education levels and locations around Australia.

The results? Health and climate issues topped the list for 32 and 30 per cent, respectively, of respondents. On the health front the most pressing issues for participants were cancer, obesity and globe-straddling pandemics. Pollution and water were the environmental issues of greatest concern and cyber-security was the principal cause for anxiety in the technology realm.

Proving that Australians really are intellectually adventurous, most respondents said that, after health, what most inspired them was space. It may be abstract and have no perceived effect on daily life but it's clear that curiosity abounds. Are we alone? How do scientists detect what's out there? What on Earth – or in space – was Einstein really talking about?

While it may come as a surprise to some headline writers, bloggers and talkback radio hosts, the OCS survey results suggest there is no widespread concern that scientists are up to no good, conducting dangerous or unethical research. Nearly half the respondents indicated they are not worried about the role science and scientists play in society, although 12 per cent expressed concern about genetic modification and 8 per cent worry about climate change. That's fair enough given the uneven quality, and quantity, of information available to the public.

All up, the survey findings reveal that Australians – like scientists – are a curious lot. The essays collected here build upon that curiosity and the recognition that if the island nation is to stay afloat, it is vital that well-trained minds wield the tools of scientific inquiry. The essays acknowledge that the most effective way to enhance Australia's social, economic, physical and intellectual wellbeing is to sort truth from fantasy, attainable dreams

from wild conjecture and apply the findings to evidence-based national, regional and local government policy... and to the Big Questions. Will ET phone home? Where did life begin? And will intelligent robots ever be able to make a good cup of coffee?

Enter *The Curious Country*. This e-book was compiled as a direct response to the survey results. It fills that critical gap between specialist publications and the increasingly limited information available about science and society through mainstream media. *The Curious Country* can be read online, downloaded to assorted electronic devices or printed for readers who like the look and feel of paper. I vote for papyrus myself.

A hand-picked group of 27 academic experts and science writers have contributed informative and entertaining essays about how

What are the scientific curiosities and concerns on the minds of Australians? What worries them, baffles them, and sets their curiosity meter to 10 out of 10?

scientists are tackling the issues highlighted by survey respondents. There's no need to read the essays in order. Dip in and out at whim.

Each essay provides a brief background to the topic and highlights the role of scientists, particularly Australian scientists, in understanding and meeting national challenges or scratching curiosity itches. Think marine science, smart buildings, cancer, the complexity of the human brain, the philosophy of science itself and the intriguing and delightfully wacky worlds of black holes, gravity waves, dark energy and other denizens of the cosmos.

To get the ball rolling, a tip of the hat must go to the scientific method. No system is perfect, but the scientific method is the most effective approach to acquiring knowledge yet devised. It beats superstition and divine revelation hands-down as ways of understanding the world.

Very simply, the scientific method is founded on the notion that nothing is true unless demonstrated to be so. A conjecture, idea or hypothesis must be tested using observation, measurement, experimentation and further testing. If a conjecture successfully predicts an outcome – say that infection with a type of bacterium will cause tuberculosis or that light will travel the same speed in a vacuum as in air – it is likely to be true. The most reliable hypotheses are called theories, and theories that consistently describe, explain and predict phenomena are scientific laws.

This sounds simple, but in practice the science process is complex. Experiments must be as free as possible from personal bias or error. Other scientists must be able to replicate the results. Hypotheses are reworked, theories

The essays collected here build upon that curiosity and the recognition that if the island nation is to stay afloat, it is vital that well-trained minds wield the tools of scientific inquiry.

blended and tested, new ways of exploring the world are developed while some hypotheses may remain untestable. As an essay in Chapter 7 on the philosophy of science illustrates, scientists themselves question what they do and how they do it.

Clearly, science is a human process. Yet unlike competing systems of knowledge, say religion or astrology, its success has cured diseases, put people into space, revealed the origins of humanity, tracked the behaviour of entities so small they may be detectable only by their effect on other entities, and revealed planets far from home. The fruits of this intellectual activity are the subject of *The Curious Country*.

Undoubtedly, one of the most contentious topics of Australian public discourse is climate change. The seas are rising! No, they're not! The planet's temperature is going up! No, it isn't!

And so it goes. Opinion-makers battle it out, leaving the public confused, frustrated and often annoyed by all the, well, hot air. Yet there is a scientific side to what is essentially a dispute about what to do – or not do – about the growing evidence that humanity's penchant for burning fossil fuels is changing the planet's weather systems.

Chapter 2, **Living in a Changing Environment,** lays out that evidence and clarifies the global and local implications of human-induced (anthropogenic) climate change. For instance, ice in Antarctica, Greenland and on mountain tops is melting, feeding a rise in sea levels. Some types of extreme weather events are likely to increase in frequency and severity. In Australia that could mean more severe floods, drought and heat events, not to mention bushfires.

Then there are other threats to the plants and creatures aboard the Australian ark. Habitat loss due to development, intensive utilisation of natural resources, the introduction of non-native species, and pollution are putting enormous pressure on the continent's flora and fauna, its so-called biodiversity.

There is growing evidence that the oceans around the ark are experiencing changes in temperature and chemistry that could lead to profound changes in the marine environment. Food webs and nutrient cycles show signs of disruption that could have impacts on economically important fisheries.

Meanwhile, there are significant problems in air pollution. Despite improvements in emission control and regulation of industry, cars and domestic pollution sources, air quality remains a major, ongoing risk to human health. Major sources of air pollutants in Australia include transport-related emissions, industrial processes such as smelting, mining and power stations, and residential sources such as domestic petrol mowers and wood heaters.

These are not welcome findings. But without research, problems are not identified and

The number and severity of extreme weather events is increasing. This satellite image from 24 January 2003, shows active bushfires as smoke billows over Victoria and New South Wales. The swirls of colour in the waters west of Tasmania could indicate a bloom of phytoplankton.

potential solutions are neither sought nor found. This is why Australian and international researchers are applying adaptive management techniques and using scientific tools to expand the understanding of the planet's changing environment. They are working to protect and nurture natural ecosystems, the bedrock of life on Earth.

Many Hollywood thrillers have depicted the terror that a mysterious and deadly disease can wreak as it spreads out of control, growing from a local then regional epidemic to a terrifying global pandemic. But the risk of dangerous, even deadly, diseases is not a Hollywood fiction. Nor are debilitating disorders such as mental illness, spinal cord injury, dementia, and the suite of life-threatening conditions, such as cancer and cystic fibrosis.

The essays in Chapter 3, **Promoting Health & Wellbeing**, reflect the key role scientific research plays in responding to the management and treatment of such diseases and conditions. The essays make sense of intriguing but often poorly understood areas such as age-related disorders and the connection between 'nature' (our genes) and 'nurture' (the environment). They sort hype from hope in the field of stem-cell research and demonstrate why advances in areas as diverse as mental illness, non-communicable diseases and vaccine development benefit from taking a society-wide look at the trends and challenges such topics pose.

Here again, scientists worldwide are in the field, the clinic and the laboratory. They are working to unravel the tangle of human and environmental factors that weave the tapestries of everyday health and evolutionary wellbeing.

Australia is the driest inhabited continent on Earth. Yet population and economic growth are driving increased demand for water, while simultaneously climate change is shifting the distribution of rainfall. There are no easy answers, but as the essays in Chapter 4, **Managing Our Food & Water Assets** attest, scientists can help Australia understand and manage the nation's food and water wealth.

Right now, researchers are teasing out the complex interconnections between water and weather dynamics. They are applying those insights to improving agricultural and food processing systems. The goal? Helping producers

The costs and benefits of genetically modified foods are yet to win over the general public, but they are a powerful tool for guaranteeing food security as climate patterns change.

deliver high quality, safe and sustainable products to tables in Australia and overseas. Their work promises to ease the transition from the mining boom to the dining boom.

But scientific innovation is not always immediately adopted by enthusiastic producers and consumers. It's not a cut-and-dried process – far from it. As the OCS survey found, some Australians are uncertain, for instance, about the costs and benefits of genetically modified (GM) food. An important essay in this chapter takes a hard look at what's what with GM food, a potentially powerful approach to food security. Another explores advances of the non-GM variety in wheat production.

If Australia is to adapt to 21st century demands for food and water, solutions must be developed and assessed. So too, must the use of Australia's biggest water tap, the Murray-Darling Basin, and the continent's underground water tank, the Great Artesian Basin. Fortunately, 21st-century scientists are on the job. Politicians, policy-makers and the public must take it from there.

It's almost laughable. An email lobs into your inbox with the exciting news that "you're the lucky winner" of a cash prize. Just forward your banking details and the cash will be deposited. You laugh and delete. The scam was obvious. But many malevolent messages are not so easy to spot. Just opening an attachment may set in train some serious cyber villainy, a topic covered in Chapter 5, **Securing Australia's Place in a Changing World**.

Contributors to the chapter also discuss the challenges and opportunities posed by globalisation of world trade and increased connectivity. They note that while science is the key tool used to develop technologies and systems capable of protecting Australians from cyber malevolence, it is also central to the creation of naval and environmental defence systems. Australia is, after all, an island nation.

It is a connected nation, too, thanks to 24/7 communications that let us trade ideas as well as products. Increasingly, science serves a diplomatic function, with international collaborations building virtual bridges between countries. Such constructive personal, national and scientific relationships will prove invaluable in securing Australia's future in a time of global uncertainty and regional change.

As science builds on itself – accepting, rejecting and refining ideas – so too does technology. And like science, technological advance is not a smooth process that begins with a Eureka moment and ends in a shiny new widget. As highlighted in an essay in Chapter 6, **Sustainable Energy & Productivity**, technological innovation is a messy, complex process, as prone to failure as success. That's why experts are seeking ways to enhance its effectiveness. This is vital. Technological innovation is the engine of success.

A case in point is the FJ Holden. While the fuel consumption of the 1950s automotive legend differs little from many similar-sized vehicles built today, engineers have not been asleep at the wheel. Modern cars are safer and less polluting than their predecessors,

From climate change to swine flu, astrophysics to zoology, this diverse group of men and women raise issues, seek answers and rigorously explore the curiosities and concerns facing today's world.

while also offering better performance and greater comfort.

But there's more to be done, and not just with cars. Australia must reduce its dependence on fossil fuels in order to limit greenhouse gas emissions and to move to sustainable, locally derived energy supplies. That means more efficient buildings as well as transport systems: planes, trains and automobiles.

It also means multiple energy options. Scientists, technologists and engineers are investigating alternative energy sources such as solar, wind, next-generation biofuels and even 'new' nuclear, as essays in this chapter illustrate. Energy-storage systems are getting a rethink as well. Clunky conventional car batteries are so last century. Today, experts talk about 'vehicle power plants', mix-and-match systems

that may run on electricity, alternative fuels or chemical reactions kicked off by novel battery components. It sounds like science fiction, but it will soon be science fact.

In March 1952, Swiss-German writer Carl Seelig received a letter from a man about whom he was writing a biography. "I have no special talent," Albert Einstein wrote in the letter. "I am only passionately curious." Personally, I'm not convinced.

Regardless, one of the greatest thinkers of all time does have a point. Humans are inherently curious. It's what we do. And thanks to a big and powerful cerebral cortex and opposable thumbs we do it better than other animals, even super-smart dolphins, chimps and Rex the Wonder Dog. It could be argued that the scientific enterprise is the ultimate expression of *Homo sapiens'* inquisitive nature. It's certainly the most effective, focused and refined reflection of it. Hence, the final chapter in this book, **Curiosity**.

The bulk of *The Curious Country* reveals the myriad ways scientists are tackling society's most pressing challenges. Nonetheless, the value of curiosity-driven research cannot be underestimated. Not only does such inquiry tickle that bulging cerebral cortex, it helps unravel mysteries that, in turn, benefit society.

True, this chapter doesn't provide the answer to the ultimate question of Life, the Universe and Everything, as posed by sci-fi writer Douglas Adams. Given that his answer was 42, it's probably just as well. But this final collection of essays will stimulate curiosity on some fascinating topics. How did the world begin? Is there life on other planets? Why do kids – just like Einstein – grasp gravity waves while adults miss the conceptual boat? Can scientists ever write the full story of the human brain, and how do scientists do science anyway?

Welcome to the Big Questions. Oh, and let me know if that artificial intelligence near Alpha Centauri makes a good espresso.

The ice shelves connecting James R Island to the continent collapsed in the 199

ANTARCTIC ICE... GOING, GOING, GONE?

As ice on the frozen continent melts, scientists examine the climate change clues captured in layers of snow that are several centuries old, writes Nerilie Abram

T WAS JANUARY 2008 and I was on the back deck of *HMS Endurance*, wearing a full-body survival suit and eager for the short helicopter ride that would take me onto Antarctic ice for the first time. The ship was travelling through the channel that divides James Ross Island from the Antarctic Peninsula – a trip that would have been impossible not so long ago.

Since the 1990s, a series of ice shelves along the Antarctic Peninsula have collapsed, including the ice shelf that had once permanently connected James Ross Island to the rest of the continent. Most famously, the collapse of the nearby Larsen B ice shelf was captured by satellite photographs. These images have been held up as an example of climate change happening before our eyes. But are they? This was what I was here to find out.

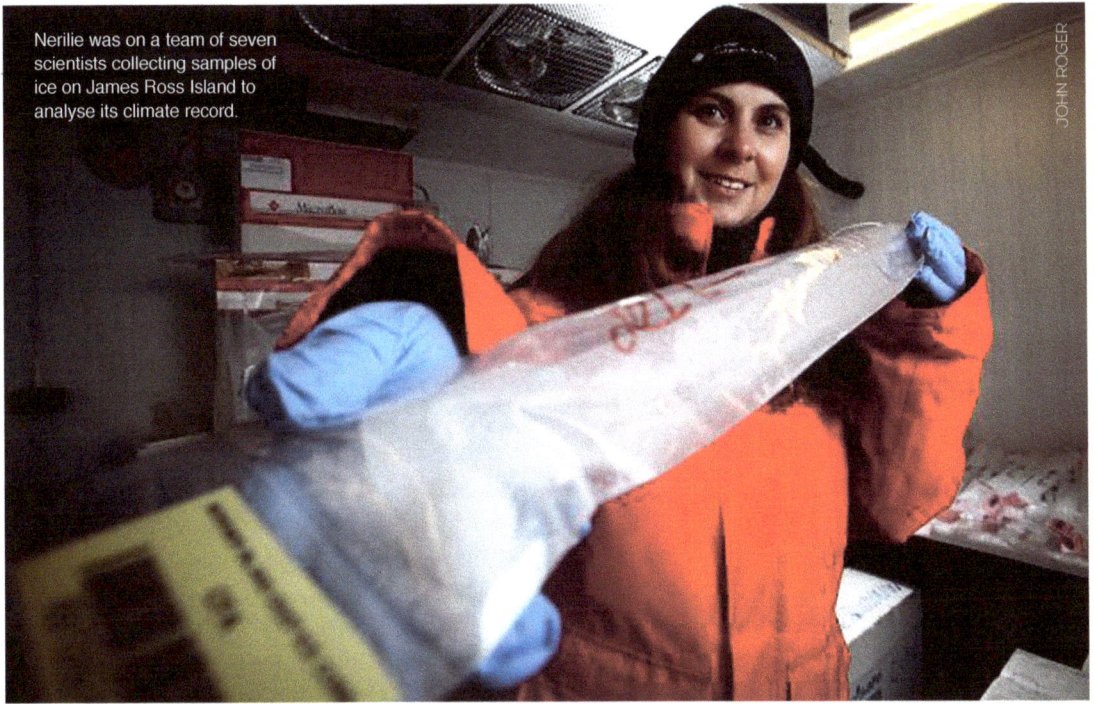

Nerilie was on a team of seven scientists collecting samples of ice on James Ross Island to analyse its climate record.

ICE AND CLIMATE

The Antarctic Peninsula is warming quickly. Over the last 50 years the climate here has warmed three times faster than the global average. The problem is that temperature measurements in this remote region don't go much further back than that. So how can we put the current warming into perspective?

The answer lies locked within Antarctica's ice. The ice blanketing most of the Antarctic continent is made of snow that has fallen and been buried. Scientists use these ancient ice layers as a window into Earth's past climate.

The deepest parts of Antarctica's great ice sheets might hold a climate record that goes back more than 1 million years. In the 2014 summer, scientists from the Australian Antarctic Division will lead an ice-drilling expedition to Aurora Basin. This is part of a coordinated international effort towards the most ambitious and technically challenging piece of ice core research ever attempted – the quest for Antarctica's 'oldest ice'.

For the much smaller James Ross Island ice-drilling project, our team of seven scientists and engineers lived and worked in tents on the ice for almost two months. The top 283 metres of this ice cap consists of snow that's built up over the past 1000 years. We know the age of the snow layers by counting the yearly summer-winter cycles of chemical impurities, such as sea salt in the ice, and by the fixed time markers left in the snow by ash from volcanic eruptions.

To build a record of how temperature changed in the past, we measure the proportion of heavy versus light water molecules, or isotopes, in the ice. Isotopes are versions of the same element that have different numbers of neutrons, so have different masses. In ice we measure the number of water molecules that have a heavy hydrogen atom (deuterium, with an atomic mass of 2) compared to those with the light hydrogen atom (atomic mass of 1). The heavy molecules take more energy to move through the water cycle, and in warm climates more of these heavy molecules will reach Antarctica and fall as snow. So the proportions of these molecules act as a 'thermometer' for the past.

The isotopes in the James Ross Island ice core tell us the coolest time on the Antarctic Peninsula was around 600 years ago. Back then the climate was around 1.6°C cooler than today. The ice also confirms that the warming here since the 1920s has been

James Ross Island is a 'Goldilocks' location for exploring the connection between temperature and ice melt. It is not so cold that summer temperatures are never high enough for melting to occur, and neither is it so warm that extensive melting destroys the climate record locked in the ice.

exceptionally fast – faster than at almost any other time in the past 1000 years.

But this particular ice core reveals much more about the changing climate on the Antarctic Peninsula. James Ross Island is a 'Goldilocks' location for exploring the connection between temperature and ice melt. It is not so cold that summer temperatures are never high enough for melting to occur, and neither is it so warm that extensive melting destroys the climate record locked in the ice. Serendipitously, conditions on this ice cap are just right for preserving a rare history of summer ice melt.

The 1.6°C of warming over the past 600 years may not sound significant, but it's caused a tenfold increase in the amount of summer melting on James Ross Island. Most of this intensification of ice melt occurred in the past few decades. This unique history of summer ice melt is a powerful illustration of how environmental changes in a warming climate don't always occur gradually.

Ice melt is an example of a threshold in Earth's environment. When summer temperatures remain below 0°C, no melting occurs. But as the climate warms towards this threshold, on some days in summer the temperature will go above 0°C and there will be excess energy to melt the surface snow. Any further warming will increase the number of days that go over the melting threshold, and increase the level they exceed it by. In this way, a small increase in average temperature can cause a large increase in melting.

So are images like the Larsen B ice shelf collapse evidence for recent climate change? Measurements from the ice core say they are.

It shows us that rising temperatures have taken summer ice melt on the Antarctic Peninsula to a level unprecedented for at least the past 1000 years. Ice melt is a critical process that weakens the structure of ice shelves and glaciers, and satellite images show that extensive summer melting caused the visually dramatic Larsen B collapse. Ice melt also has real implications for rising sea levels across the world.

ANTARCTICA'S ROLE IN GLOBAL SEA LEVEL RISE

Rising sea levels in a warming world is particularly relevant to Australia as large proportions of our population and infrastructure are near the coast.

In 2013, the Intergovernmental Panel for Climate Change (IPCC) released its fifth assessment report. On our current emissions trajectory it projects that sea level is likely to rise by between 0.53 and 0.97 metres by 2100. This projection takes into account the thermal expansion of the oceans as they warm, as well as changes in snowfall, surface melting and glacier loss that will alter the quantity of ice locked up on land. What these model-based projections aren't yet able to assess is the possibility of accelerating ice flow and loss from Antarctica's vast ice sheets.

Antarctica's contribution to sea level is a balancing act between ice accumulation across the central plateaus and ice loss around the margins of the continent. Satellite monitoring of Antarctica's ice sheets over the past few decades has revolutionised our understanding of this changing balance. These satellite measurements use changes

For two months, large weather-haven tents housed the team's living, working and ice-drilling areas.

in the height or gravitational pull of the ice sheets to identify places where Antarctica is gaining or losing ice.

Overall, Antarctica is losing ice, accounting for just under 10 per cent of the rise in global sea level over the past two decades. The mountain glaciers and ice caps along the Antarctic Peninsula are losing around 20 billion tonnes of ice yearly. Even more significant is the approximately 65 billion tonnes of ice lost each year from West Antarctica.

This is just the tip of the iceberg, so to speak. West Antarctica has been described as the 'weak underbelly' of Antarctica's ice sheets. This ice sheet sits on bedrock, but that ground is below sea level – by more than 2 kilometres in some places. This makes the ice sheet vulnerable to melting from beneath. As the margins of the West Antarctic Ice Sheet melt and thin,

seawater warm enough to melt the ice is able to encroach further under the ice sheet, and this could cause ice to be lost even faster. The latest IPCC report flags the possibility that rapid collapse of parts of the West Antarctic Ice Sheet could cause sea level to rise substantially above its current projections.

Earth's past provides some evidence to gauge how quickly ice could be lost from Antarctica in the future.

The last time the Earth's temperature was similar to today – around 125 000 years ago – sea level was roughly 6 metres higher and changed twice as fast as the sea level rise we've seen in the past decade. What this demonstrates is the ability for sea level to respond to climate warming at a speed that matches the upper end of IPCC projections for the 21st century.

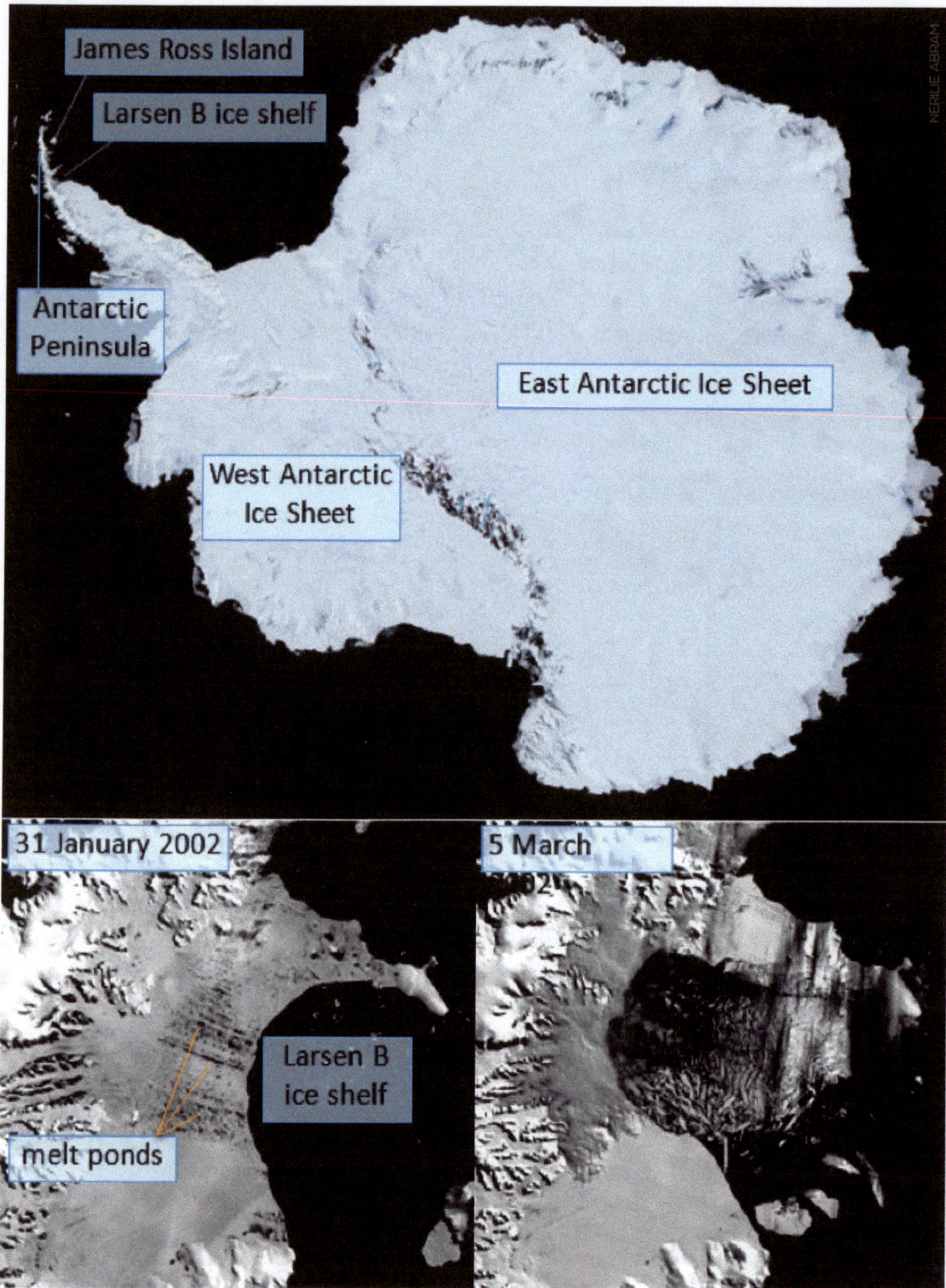

The frozen continent: The locations discussed in this essay are shown on a satellite image mosaic of Antarctica. The lower panels show satellite images captured during the Larsen B ice shelf collapse in 2002. Extensive melt ponds were visible on the ice shelf in late January, which weakened its structure and, over the space of just a few weeks, more than 3200 square kilometres of the ice shelf disintegrated. The nearby Larsen A and Prince Gustav ice shelves were also lost in the decade before the Larsen B collapse.

ALL THINGS ICE

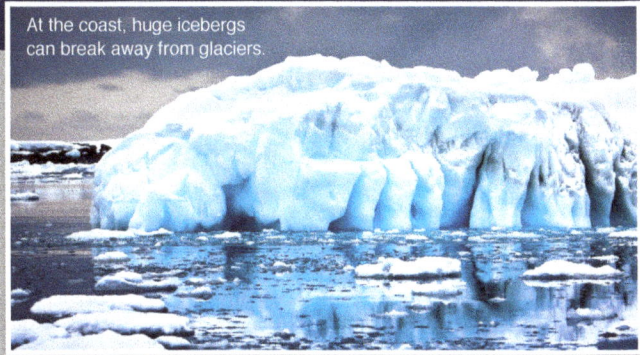

At the coast, huge icebergs can break away from glaciers.

Ice sheets are large domes of land ice. The East Antarctic Ice Sheet occupies over 75 per cent of the continent and holds the equivalent of 53 metres of sea level. The smaller but less stable West Antarctic Ice Sheet has the equivalent of 4.3 metres of sea level locked within it. Ice caps are small versions of ice sheets.

Glaciers are like rivers of flowing ice. They are found on the edges of Antarctica's ice sheets and along the mountainous Antarctic Peninsula. At the coast, the ice from glaciers can break away as icebergs, or feed into a floating ice shelf.

Ice shelves are where Antarctica's land ice extends out over the sea. Because ice shelves are floating, their ice has already contributed to global sea level. Ice shelves act as a buttress for the glaciers behind them. When they are lost the flow of glacier ice into the ocean speeds up. Around 28 000 square kilometres of ice shelf area around Antarctica has been lost since the 1950s.

Sea ice forms when the ocean's surface freezes. In winter, the sea ice around Antarctica effectively doubles the size of the continent. Sea ice doesn't have any influence on sea level, but it is important for climate through the exchange of carbon dioxide between the ocean and the atmosphere and because it is white, the surface absorbs less heat than the dark ocean.

Sea level in the past closely followed the changes in polar temperatures recorded by ice cores. This connection provides another way to determine the possible trajectory of future sea level rise. There are uncertainties in this approach, but the observed relationship between temperature and sea level since the 1880s indicates that the IPCC's estimates for future sea level rise may be too low. Projected warming for the coming century points to the possibility that sea level could rise by as much as 1.9 metres by 2100.

The potential for rapid changes in ice melt and loss in Antarctica presents an enormous challenge for Australia's efforts to plan adequately for rising sea levels. Scientists will continue to unlock the clues that Antarctica's vast ice sheets contain about her past. This will provide a long-term context that is critical to our understanding of the changes we are now seeing – and those that lie ahead for Antarctic ice.

DR NERILIE ABRAM is a palaeoclimate scientist. She holds an Australian Research Council QEII Fellowship at The Australian National University. Her ice core research is part of an international collaboration with the British Antarctic Survey.

FURTHER READING

Scientific Committee on Antarctic Research 2009, *Antarctic Climate Change and the Environment*, SCAR, Cambridge, UK.

Australian Antarctic Division 2008, 'Ice sheets and sea-level rise', http://www.antarctica.gov.au/about-antarctica/environment/climate-change/ice-sheets-and-sea-level-rise.

Vermeer, M. and Rahmstorf, S. 2009, 'Global sea level linked to global temperature', *Proceedings of the National Academy of Sciences of the United States of America* 106(51): 21461-21462.

Abram, N.J. et al 2013, 'Acceleration of snow melt in an Antarctic Peninsula ice core during the 20th century', *Nature Geoscience* 6, 404-411.

Grant, K.M. et al 2012, 'Rapid coupling between ice volume and polar temperature over the past 150 000 years', *Nature* 491, 744-747.

MARINE LIFE IN A CHANGING CLIMATE

Rising greenhouse gas emissions are warming the marine environment with unprecedented consequences, writes Elvira Poloczanska

WE LIVE IN a naturally variable world. The climate system contains phenomena such as El Niño-Southern Oscillation and the Indian Ocean Dipole that affect Australia's climate and beyond, over time-scales that range from years to many decades. (See: **See-saws and boys and girls**).

But we are heading into the unknown. Increasing levels of greenhouse gases in the atmosphere are associated with warming in the climate system, including both the atmosphere and ocean. Insidious and pervasive changes to components of the climate system are unprecedented, and in this case over time-scales ranging from decades to hundreds of thousands of years. Land and water temperatures are increasing, sea levels are rising, wind patterns and ocean currents are changing and the oceans are acidifying (See: **Ocean acidification**). The natural world is in continuous ebb and flow in response to such climate variability. So why is climate change any different?

For a start, not only are human-induced, 'anthropogenic', greenhouse gas emissions leading to changes unprecedented in the historical record, but the rate of many of these changes will be rapid, possibly too rapid for some components of the natural world to adjust.

Marine life, such as coral polyps, is being affected by anthropogenic climate change.

Impacts of recent climate change have been observed via the responses of hundreds of individual species and fundamental ecosystem processes. An international team, led by me and Commonwealth Scientific and Industrial Research Organisation (CSIRO) colleague Anthony Richardson, has pooled information from many studies and all of the oceans to show a global fingerprint of the impact of recent climate change on ocean life. From microscopic plankton floating at the ocean surface, to sea turtles migrating across entire ocean basins, marine life is being affected by anthropogenic climate change.

In Australian oceans, tropical and subtropical species of fish, molluscs and plankton are shifting southward as waters warm, while cool-water seaweeds are in decline on both sides of the continent. Mass coral bleaching has occurred regularly on the Great Barrier Reef

SEE-SAWS AND BOYS AND GIRLS

The see-sawing of sea surface temperatures between the eastern and western Indian Ocean is known as the Indian Ocean Dipole. A 'positive' dipole phase is associated with cooler than normal sea temperatures north-west of Australia and the eastern tropical Indian Ocean, and a decrease in rainfall over parts of central and southern Australia. A 'negative' phase produces warmer than normal waters off Australia's north-west coast and an increase in rainfall over parts of southern Australia.

The El Niño-Southern Oscillation (ENSO) is the dominant mode of year-to-year climate variability observed globally. ENSO oscillates between 'El Niño' (the boy) and 'La Niña' (the girl) conditions. El Niño refers to extensive warming of the eastern tropical Pacific Ocean which leads to a major shift in weather patterns across the Pacific. El Niño is associated with warmer than normal sea temperatures during late summer off northern and eastern tropical Australia and a decrease in winter, spring, and summer rainfall over much of eastern Australia.

The environmental changes are possibly too rapid for some components of the natural world to adjust.

since the early 1980s; yet no evidence of mass coral bleaching was reported in the scientific literature prior to this time. Australian scientists have worked with data from scientific surveys, statistical and mathematical models and information supplied by fishers, divers and citizen scientists (see: **Travels of a big barnacle and little snails**) to provide conclusive evidence of these recent impacts.

Further changes to the climate, and therefore the natural world, are highly likely to continue. It's time to focus on actions.

ADAPTING TO CLIMATE CHANGE

In order to prosper in a changing climate, industries and societies must be agile and adaptable. The process of adapting to climate change is already underway in Australia.

Scientists are producing much-needed information to assist ocean managers and policy makers to help marine ecosystems and industries adapt. For example, fishery managers may need to consider management changes as fish species move across state boundaries or as stock productivity changes. Decisions will be informed by a variety of research methods, including regional climate models, fisheries assessment models, biodiversity surveys, habitat mapping, and even whole ecosystem models such as 'Atlantis'. Developed by the CSIRO's Beth Fulton, Atlantis considers biophysical, economic and social components of marine ecosystems.

With adaptation planning we can capitalise on opportunities and reduce losses; however, we are likely to face ecological 'surprises'. For instance,

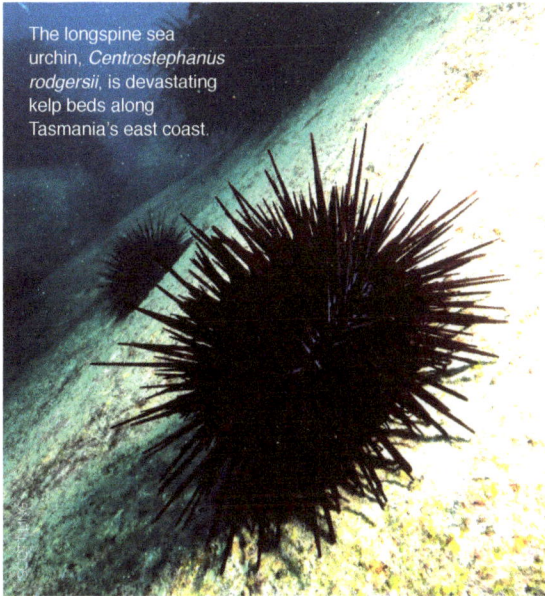
The longspine sea urchin, *Centrostephanus rodgersii*, is devastating kelp beds along Tasmania's east coast.

OCEAN ACIDIFICATION

Atmospheric carbon dioxide, released by human activities, is entering the world's oceans, making them more acidic. There is growing evidence that even relatively minor perturbations in ocean chemistry could lead to profound changes in the marine environment. We only have to look at the research of CSIRO, Australian Institute of Marine Science, University of Queensland, James Cook University, and the Antarctic Climate and Ecosystems Cooperative Research Centre (CRC) to understand that the whole ocean, from tropics to poles, is acidifying.

Ocean acidification not only affects the ability of corals and other animals and plants to form shells and skeletons of calcium carbonate, it can also disrupt basic physiological and behavioural processes of fish.

We have very sophisticated approaches to examine the impacts of ocean acidification in Australia. These include global and regional ocean models, mini coral reefs in experimental tanks, and aquaria for rearing larval fish. Scientists have also undertaken research expeditions to natural carbon dioxide seeps found in shallow waters off Papua New Guinea, where the gas bubbles from the ocean floor due to volcanic activity. These seeps give us a glimpse of coral reefs in an acidified ocean, and such reefs are not the rich and beautiful complex of thousands of coral species we see elsewhere today, even just a few kilometres from the seeps. Instead these acidified reefs are dominated by a few slow-growing species and are not able to support the high diversity of life we see on coral reefs today.

having species of tropical fish turning up further south may be good news for recreational fishermen and divers, who will have new species to 'bag', but such range shifts may also bring challenges.

Case in point: the grazing urchin, *Centrostephanus rodgersii* (pictured above), has greatly multiplied in numbers and spread down the Tasmanian coastline in response to warming and a strengthening of the East Australian Current. It has begun to change the structure of Tasmanian kelp communities with detrimental effects on coastal biodiversity including commercial species such as abalone and lobster.

Researchers Scott Ling and Craig Johnson, with the Institute for Marine and Antarctic Studies in Tasmania, showed that over-fishing of large lobsters, one of the urchin's chief predators, has helped urchins to overgraze important kelp-bed habitat. Adapting fishery management to enhance stocks of larger lobsters could be a win-win situation, benefiting both the fishery and the kelp ecosystem.

Australia is a land of extremes. It suffers floods and heatwaves, summer cyclones batter tropical coasts and winter storms drench southern coasts. Such extreme events give

a glimpse of life in a warmer world and allow researchers to develop and test adaptation responses. In the summer of 2010-11, an unusual 'marine heatwave' occurred in the coastal waters off central Western Australia. An unusual coincidence of climatic events led to water temperatures 5°C above normal, which lasted for some ten weeks.

The effects were profound. Widespread 'die-offs' of fish, seaweeds and shellfish occurred and tropical species moved farther south temporarily. Whale sharks and manta rays were

TRAVELS OF A BIG BARNACLE AND LITTLE SNAILS

Clues to the impacts of climate change far offshore can be discovered closer to home, along Australia's rocky coastline. Scientists comprehensively surveyed shores in Queensland, New South Wales, Victoria, South Australia and Tasmania during the 1950s and 1960s, a stretch of some 4000 kilometres and a major undertaking at the time. These extensive broad-scale surveys mapped the distributions of intertidal plants and animals and give a baseline against which to compare shores today.

We recently revisited many of the historical survey sites in south-eastern Australia and found shifts in the distribution of some intertidal barnacles and snails consistent with the rapid warming in the region. The most striking shift was that of the giant rock barnacle *Austromegabalanus nigrescens*. We're confident the barnacle was absent from Tasmania in the 1950s as the scientists made a special effort to look for it and noted in their field books that they were 'very surprised' they couldn't find it anywhere in Tasmania.

Today the barnacle occurs from Eddystone Point in the extreme north-east of Tasmania down to the Tasman peninsula. Observations of the barnacle in Tasmania exist for the north-eastern coast from the late 1980s, so it must have arrived in Tasmania sometime before this.

We need your help to chart the changes taking place in Australia. Join the ClimateWatch program and report the species you find when you visit the coast or go for a walk. Information can be either entered on the website or by downloading the ClimateWatch app.

By getting involved, you can help scientists understand how climate change is impacting Australian ecosystems.

sighted far south of their usual range. Several fisheries were negatively impacted. Adaptation response plans were put in place to maximise recovery. The lessons learned from this, and other such extreme events can help fisheries and conservation managers plan for future climate change impacts.

ADAPTATION RESEARCH

Scientists use monitoring to underpin adaptation research. The Australian Integrated Marine Observing System (IMOS) monitors boundary currents and continental shelf waters for changes in their physical (temperature, salinity, and currents), chemical (nutrients and carbon) and biological (plankton and top predators) characteristics using state-of-the-art technology such as ocean gliders (see picture), moorings, floats and satellite imagery. This national observation system is advancing understanding of climate variability and change, and providing an early warning system for major changes that might occur.

At the CSIRO Climate Adaptation Flagship, teams of climate scientists, conservation

We must keep our focus on reducing anthropogenic greenhouse gases in the atmosphere.

biologists, ecologists, social scientists, engineers and economists work together to develop and inform adaptation approaches. Adaptation is not just about technological and engineering approaches, but also about knowledge delivery and education. These include delivering bespoke seasonal forecasts to aquaculture facilities along coastlines, delivering future scenarios of climate and natural systems to policy-makers, and synthesising observed and expected impacts for the public and decision-makers.

Climate change is only one pressure on our natural world but it will force us to explore new pathways and to think 'outside the box' as we face new combinations of environmental conditions. For example, we may need to breed new aquaculture species that are suited to future climates, to translocate vulnerable plants and animals to new locations where they can survive, to build safety margins into fishery harvest levels to account for uncertainty in climate change impacts, and to remove hard barriers, such as walls and buildings, to allow

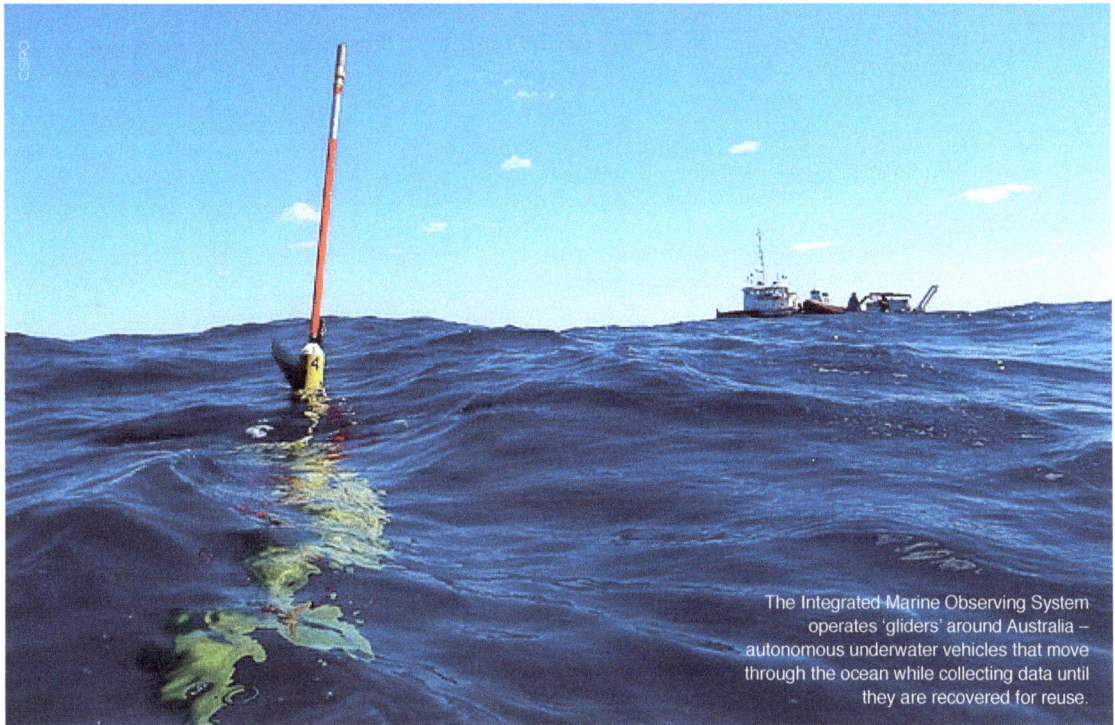

The Integrated Marine Observing System operates 'gliders' around Australia – autonomous underwater vehicles that move through the ocean while collecting data until they are recovered for reuse.

mangroves and salt marshes to retreat as sea levels rise. Understanding the genetic basis of adaptation within species will help us to understand their vulnerabilities to climate change and allow us to ensure that gene combinations able to can confer resilience to future climate change prevail.

BEYOND ADAPTATION

Investment in adaptation planning now can fortify us for the inevitable changes to come, but we still need to look at the big picture. There may be thresholds beyond which we can't adapt, and unexpected challenges to face.

We must not take our focus off our primary goal, which should always be mitigation. If we want our children and our children's children to enjoy a rich and varied natural world, to have food security and enjoy the privileges of the technology we enjoy, we must keep our focus on reducing anthropogenic greenhouse gases in the atmosphere.

DR ELVIRA POLOCZANSKA is a research scientist at the CSIRO Marine and Atmospheric Research Division. She specialises in climate change ecology, ecological modelling and coastal ecosystems.

FURTHER READING

CSIRO Climate Adaptation Flagship, website, http://www.csiro.au/en/Organisation-Structure/Flagships/Climate-Adaptation-Flagship.aspx.

Poloczanska, E.S., Hobday A.J. and Richardson A.J. (eds) 2012, *Marine Climate Change in Australia Impacts & Adaptation Report Card 2012*, ebook, http://www.oceanclimatechange.org.au/content/index.php/2012/home/.

ClimateWatch, website, http://www.climatewatch.org.au/.

Poloczanska, E.S. et al 2013, 'Global imprint of climate change on marine life', *Nature Climate Change* 3(10), 919-925 (or read a summary at *The Conversation*).

Atlantis, ecosystem model developed by researchers at the CSIRO, http://atlantis.cmar.csiro.au/.

Bushfire smoke towers above a suburban beach near Hobart on 4 January 2013 – a day of record-breaking fire weather.

ADAPTATION IS THE KEY TO SURVIVAL

Raging bushfires, torrential rain, floods and droughts are warnings that it's time for Australia to change its ways, writes David Bowman

RECENTLY, I VISITED the Australian Alps – the landscape that inspired me to become a landscape ecologist. My boyhood memories from 40 years ago were confronted by a fire-blasted landscape. More than 90 per cent of the Australian Alps bioregion has been burnt in a series of massive bushfires since the start of this century. At my old school, nestled in the foothills of the Alps, I discovered a state-of-the-art fire bunker with its own water, power and air supply designed to provide refuge in case of another extreme bushfire.

This experience highlights the fact that when people are confronted by a direct environmental threat they change behaviour. They adapt. But adaptations can be economically costly, socially disruptive and in some cases environmentally damaging, so there must be clear and compelling evidence that what has worked in the past will not work in the future. This is a job for science and scientists. We can – and should – play a pivotal role in determining when, how and why Australia responds to environmental change.

MORE SEVERE BUSHFIRES?

Extreme environmental changes, such as bushfires, have always occurred because we're on a dynamic, living planet. What is at stake is whether our society and economy can persist if the *rate* of environmental changes abruptly increases. For instance, a massive surge in bushfires worldwide could, in principle, accelerate climate change by releasing massive stores of carbon dioxide (CO_2) and soot, which will lead to warmer and drier climates favouring yet more fires. Even a moderate increase in fire activity will provide a brake on Australia's economy, given the economic and social impacts of these disasters.

Determining whether bushfires have increased in severity would be possible if perfect historical records existed, stretching back thousands of years. Unfortunately, in Australia there are very few detailed historical records of natural phenomena or variation in the geographic patterns of plant and animal populations.

To build a picture of past environmental changes, scientists use 'environmental archives' such as sediments in lakes, tree rings, and historical records of landscapes made by painters, photographers, and explorers. These data provide a baseline to determine if current environmental conditions are outside the range of natural variability.

Currently in Australia, the historical data are too sparse to answer the question unambiguously of whether bushfire activity is within or outside the range of historical variability. Nor is there enough evidence to determine if climate change or the cessation of Aboriginal landscape burning is causing bigger and more intense fires; only through increased investment in scientific research can these issues be resolved.

Satellites, airborne sensors and computer networks are creating a revolution in the capacity to record bushfires across landscapes. These monitoring systems provide essential information about fire activity which can be analysed to understand better the success and failure of fire management approaches and the effects of fire on plants and animals.

HUMAN ADAPTABILITY

Humans have a remarkable ability to adapt to threats. It's what our ancestral *Homo sapiens* did for hundreds of thousands of years and it's what we're doing today.

For instance, there's growing awareness among people living in flammable environments about the need to fireproof homes and have gardens with non-flammable plants. More planned burning is being carried out to reduce fire hazards.

Because one size does not fit all, scientists must discover how to achieve sustainable co-existence with bushfires. For specific ecosystems, we need to know what is true and what is not, what is practical and cost effective and what is not.

For example, the heated debate about whether cattle grazing in the Australian ecosystems reduce bushfires is based on slender evidence and is context-specific.

In certain types of ecosystems, such as semi-arid rangelands, grazing may reduce fire severity by controlling rampant invasive non-native grass. These grasses can fuel fires of sufficient intensity to kill trees, degrading wildlife habitat. But in other ecosystems, such as the Australian Alps, evidence shows grazing does not reduce bushfire severity. Instead, it causes substantial ecological damage, such as trampling fragile bogs and fouling creek lines.

ADAPTIVE MANAGEMENT

Adaptation to environmental change can involve trial and error to find solutions to problems. Science, using logic and experimentation, has a vital role in fast-tracking this process. Case in point: prescribed burning. Scientific analyses

Since the beginning of the 21st century, high-severity fires have killed large areas of tall eucalypt forests in the Australian Alps.

show the real benefit of planned burning in protecting lives and property comes from burning bushland close to houses. Yet research also shows such burning causes smoke pollution which can harm health, albeit less than that from intense bushfires. By combining this knowledge it's possible to determine the risk-reduction benefit of planned burning while minimising exposure of vulnerable people to smoke.

A cycle of clarifying problems, designing interventions and monitoring the outcome is often called 'adaptive management'. For instance, as a consequence of the recommendation of the 2009 Bushfire Royal Commission, Victoria is almost tripling the amount of planned burning across the state. But in addition to this increase the Royal Commission also recommended monitoring and evaluating the impact of the burning to understand the effect on biodiversity and the environment.

The application of adaptive management will become increasingly important as we humans are forced to respond to the multiple challenges of changing environmental conditions, increasing human populations and strains on ecosystems globally.

ADAPTATION, SCIENCE AND GLOBAL ENVIRONMENTAL CHANGE

Global environmental change is the signature tune of the 21st century. Indeed, some scientists suggest we should recognise the planet has shifted into a new geological epoch, one they call the Anthropocene.

This is the case because humans have become major actors in the functioning of the Earth system – be it by accelerating geological processes, modifying local, regional and global climate patterns, restructuring ecosystems through extinctions and introducing plants and animals, or by modification of landscapes through damming rivers, clearing forests, cultivating grasslands and mining ore deposits.

Science and technology are crucial to mitigating the effects of these changes. Already on the table are confronting options such as geoengineering – large-scale intervention in Earth's climate system to reduce global warming – and new ecosystem construction using novel combinations of plants and animals and modified species rejigged by biotechnology for conservation or adaptation. Such options raise profound ethical questions, extending beyond the reach of science alone.

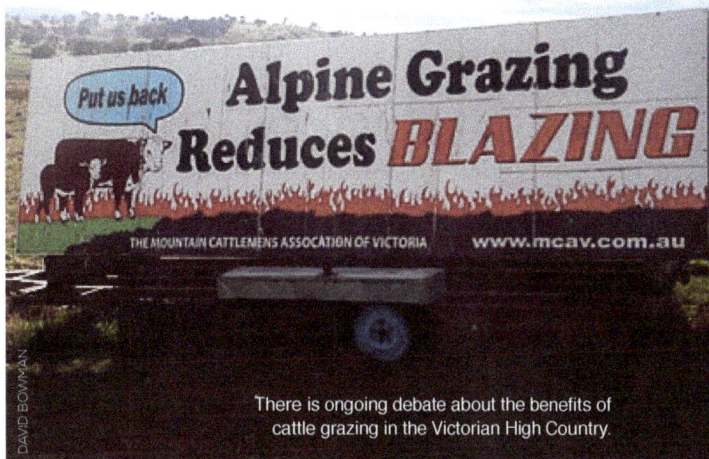

There is ongoing debate about the benefits of cattle grazing in the Victorian High Country.

I owe my career to an inspiring school teacher who opened my eyes to the power of science to help us understand the natural environment. I was lucky. Over the past 40 years, numerous scientific discoveries have changed the way Australians think and feel about the bush. Among the important contributions made by scientists has been the recognition that, along with drought and floods, bushfires have shaped our "wide brown land". We must coexist with these powerful natural forces, and science provides the key to this urgent adaptive process.

SMOKE

It's obvious. Where there's smoke there's fire; where there's fire there's smoke; and where there's smoke there are respiratory and other health disorders. But surprisingly, until a decade ago, smoke from bushfires was seen as a minor nuisance rather than a substantial health risk.

To manage bushfire smoke better, a research program is being conducted that combines studies of human health, landscape ecology and atmospheric effects. Funded by the Australian Research Council, the project, which began in 2001, includes representatives from land management, environmental protection and health departments in the Northern Territory, Tasmania, New South Wales and Western Australia.

Already, results clearly demonstrate that smoke produced from severe and even low-level bushfires can adversely affect human health. People at highest risk are the very old or the very young, or those who live with chronic medical conditions such as asthma or heart disease.

The project has contributed to improved guidelines for land management, environment protection and public health practice. For example, in Tasmania a network of smoke-monitoring sites was established and its real-time pollution readings are linked to health advice and information for the general public.

This example shows how science can help the community adapt to bushfire smoke pollution, and tolerate the effects of planned burns designed to reduce the risk to life and property from uncontrolled bushfires. Given the many challenges faced by those in bushfire regions – including urban development adjacent to bush land, hotter and drier climates, and the need to reduce smoke pollution further – it is vital that scientific research efforts are sustained.

PROFESSOR DAVID BOWMAN is a University of Tasmania landscape ecologist and bushfire scientist committed to using science to improve land management and shape conservation policy debates.

FURTHER READING

Johnston, F. 2009, 'Bushfires and human health in a changing environment', *Australian Family Physician* 38(9): 720–724.

Bowman, D.M.J.S., O'Brien, J., Goldammer, J. 2013, 'Pyrogeography and the global quest for sustainable fire management', *Annual Review of Environment and Resources* 38: 57–80.

Bowman, D.M.J.S. 2013, 'Winning the climate debate by adapting', *The Conversation*, http://theconversation.com/winning-the-climate-debate-by-adapting-12409.

'Firestorm', *The Guardian*, http://www.theguardian.com/world/interactive/2013/may/26/firestorm-bushfire-dunalley-holmes-family.

Department of Health, Tasmania, 'Wood smoke, air quality, your health', http://www.dhhs.tas.gov.au/peh/alerts/air.

Lunt, I. 2012, 'Can livestock grazing benefit biodiversity?' *The Conversation*, http://theconversation.com/can-livestock-grazing-benefit-biodiversity-10789.

The two species of
Australian native
corroboree frogs are
critically endangered.

BIOWEALTH:
ALL CREATURES
GREAT AND SMALL

**Australia's financial wealth, health and wellbeing depend on its biowealth
— the diversity of species that support all life, writes Corey Bradshaw**

AS I STEPPED off the helicopter's pontoon and into the swamp's chest-deep, tepid and opaque water, I experienced for the first time what it must feel like to be some other life form's dinner. As the helicopter flittered away, the last vestiges of that protective blanket of human technological innovation flew away with it.

Two other similarly susceptible, hairless, clawless and fangless *Homo sapiens* and I were now in the middle of one of the Northern Territory's largest swamps at the height of the crocodile-nesting season. We were there to collect crocodile eggs for a local crocodile farm that, ironically, has assisted the amazing recovery of the species since its near-extinction in the 1960s. Removing the commercial incentive to hunt wild crocodiles by flooding the international market with scar-free, farmed skins gave the dwindling population a chance to recover.

Conservation scientists like me rejoice at these rare recoveries, while many of our fellow humans ponder why we want to encourage the proliferation of animals that can easily kill and eat us. The problem is, once people put a value on a species, it is usually consigned to one of two states. It either flourishes as do domestic crops, dogs, cats and livestock, or dwindles towards or to extinction. Consider bison, passenger pigeons, crocodiles and caviar sturgeon.

The extinction of the Tasmanian tiger is an example of humans contributing to a loss of biodiversity.

TMAG

As a conservation scientist, it's my job not only to document these declines, but to find ways to prevent them. Through careful measurement and experiments, we provide evidence to support smart policy decisions on land and in the sea. We advise on the best way to protect species in reserves, inform hunters and fishers on how to avoid over-harvesting, and demonstrate the ways in which humans benefit from maintaining healthy ecosystems.

Homo sapiens is a relatively new addition to the global species pool collectively called 'biodiversity'. Like other species and physical processes before us, we have changed our planet's biosphere in a geological heartbeat. Many geologists argue that the planet has entered a new geological era – the Anthropocene – which is characterised by the human-caused signal of mass extinction above the normal rate at which species vanish.

Extinction generally comes in waves – so-called mass extinction events. Prior to the Anthropocene, five mass extinction events have occurred since the Cambrian period about 500 million years ago. The Permian extinction (250 million years ago) was the worst. Roughly 95 per cent of all species on Earth disappeared. The most infamous mass extinction happened about 65 million years ago during the Cretaceous period when a giant asteroid struck Earth, killing off most dinosaurs.

But the Anthropocene shows extinction rates exceeding the background rate – the rate between mass events – by up to 10 000 times. Of course, scientists debate the true inflation factor due to the difficulty of observing extinctions. (See: **Counting species one by one**). Regardless, it's clear the planet is losing biodiversity at an alarming rate.

Given the realities of daily life, it's easy to forget that biodiversity is important to our wellbeing. Australians feel they are in touch with the bush, but the fact is most do not appreciate the natural world on which they utterly depend.

It's not hyperbole, naïveté or green platitudes – all people depend absolutely on every other species. For instance, consider the very air we breathe. Nearly all the oxygen in the atmosphere is produced by plants and much of that by marine algae. Yet worldwide we treat oceans like giant toilets and cut down forest blocks every year that, together, equal the size of Tasmania.

On the topic of plant respiration – the process of photosynthesis in which plants take up carbon dioxide and release oxygen – the world is now faced with centuries of tumultuous climate disruption from industrial emissions, yet more than a third of the world's carbon is stored in forests. In other words, more forests equals less carbon dioxide in the atmosphere and slower, less intense climate change.

Much of the food grown to feed the seven billion-strong human population is pollinated by a wide array of animals, and most of that is done by a single species – the honeybee. Yet bee populations around the world are crashing because of forest fragmentation and our overuse of pesticides. No pollination, fewer crops. And most of the world's drinking water comes mainly from natural waterways and wetlands that filter out the contaminants people produce.

Other examples of 'ecosystem services' abound. Even the much-maligned shark is an essential ecosystem engineer. Wherever shark populations are abundant and diverse, reefs are healthier, fish populations are higher and water clarity is better.

This happens because large sharks impose a top-down pressure on smaller predators, thus limiting the latter's intake of other fish species. Removing the biggest predators means that smaller predators increase, which then quickly eat other species that keep things like algae in check. The overall effect is a biologically poor system, prone to further degradation.

Even the feared dingo plays an essential ecosystem role. Wherever scientists have looked, areas with large dingo populations have more native marsupials. Where dingos are poisoned or fenced out, native mammals do not do well.

Why? Dingoes outcompete and kill introduced cats and foxes. Australia's estimated 18 million feral cats, in particular, are a biodiversity scourge. To illustrate, imagine a line of stock trucks bumper-to-bumper along the 600 kilometres from Sydney to Grafton. Each is filled to the brim with native animals: possums, bandicoots, penguins, lizards, skinks and so forth. This represents how many native animals are killed *each year* by feral cats. Little wonder then that Australia has the world's worst record for mammal extinctions.

If one considers the totality of all these different interactions, dependencies and functions – the scientific discipline of

COUNTING SPECIES ONE BY ONE

It is easy to be impressed when considering the variety of life on Earth, known collectively as biodiversity. Conservative estimates place the number of species in different groups living today at more than 4 million protists – microorganisms without a cell nucleus – 75 000-300 000 helminth (worm) parasites; 1.5 million fungi, 320 000 plants, 4-6 million arthropods (insects and the like), 30 000 fishes, 6500 amphibians, 10 000 reptiles, 10 000 birds and around 5000 mammals.

While scientists are confident they have inventoried most of the larger species, such as mammals and birds, estimates of the number of smaller, more cryptic species are highly uncertain. In fact, total estimates range from only several million to several hundred million species worldwide. Both extremes seem unlikely.

The term biodiversity itself is a variable concept. The simplest way of estimating it is to count the number of species within a given area. But this belies its complexity. Biodiversity includes, among many other things, genetic diversity, ecological function, and the way in which species' composition changes over space and time. Simply adding up the number of species, therefore, ignores important factors like 'endemism' – species found nowhere else – rarity, genetic variation, resilience, and evolutionary potential, and the ability to adapt to environmental change by evolving. It's hardly surprising that people often have difficulty grasping the importance and complexity of biodiversity, especially considering our increasingly nature-disconnected lifestyles.

Another important aspect of biodiversity is how much of it is disappearing, and at what rate. Extinction might appear obvious because it ultimately involves comparing a time when a species was present to another when it is no longer. Unfortunately, it's not that straightforward. Even the date of the infamous dodo extinction is uncertain, with claims it survived another 30 years beyond its last sighting.

The problem lies in the fact that as population abundance declines it is more difficult to detect remaining individuals, especially in the case of already rare and cryptic species. For example, would anyone even notice if a rare species of underground fungus went extinct? The answer is: only if someone had already been documenting its distribution and decline. Expand that to the millions of species on the planet, combined with the uncertainty associated with that number itself, and it becomes clear why estimates of extinction rates are highly uncertain.

Most of the world's pollination is done by a single species – the honeybee. And bee populations are crashing.

ecology – the logical conclusion is that all biodiversity can be considered under the umbrella of 'biowealth'.

This concept encapsulates the two most important elements of biodiversity from a human perspective. The first is that diversity is an essential requirement for life. Without all, or at least most, of these species, we humans inevitably lose important services. Secondly, this diversity provides humanity – largely free of charge – with the elements essential for survival. Without biodiversity we are poor. With it we are 'biorich'.

So consider the crocodiles, sharks and snakes, the small and the squirmy, the smelly, slimy and scaly. Consider the fanged and the hairy, the ugly and the cute alike. The more we degrade this astonishing diversity of evolved life and all its interactions on our only home, the more we expose ourselves to the ravages of a universe that is inherently hostile to life.

It is time to embrace, protect and cherish Australia's biowealth so our children can live happy, prosperous lives. It is time to build biodiversity into daily life by regularly reporting the state of the nation's biowealth alongside economic, sport and stock market indices. Only then will society be cognisant of, and perhaps stimulated to improve, the state of *Homo sapiens*' one and only life-support system.

PROFESSOR COREY BRADSHAW is Director of Ecological Modelling at The University of Adelaide's Environment Institute. He specialises in applying mathematical approaches to measure the effect of humans on biodiversity and investigates the ways in which a prosperous human society can co-exist and benefit from healthy, natural ecosystems. He is a leading authority on biodiversity, with over 200 scientific publications and a popular conservation blog (ConservationBytes.com).

FURTHER READING

Bradshaw, C.J.A., Sodhi, N.S., Brook, B.W. 2009, 'Tropical turmoil – a biodiversity tragedy in progress', *Frontiers in Ecology and the Environment* 7: 79-87.

Bradshaw C.J.A. 2012, 'Little left to lose: deforestation and forest degradation in Australia since European colonisation', *Journal of Plant Ecology* 5: 109-120.

Daily G.C. 1997, *Nature's Services: Societal Dependence on Natural Ecosystems*, Island Press, Washington, D.C.

Sodhi, N.S., Brook, B.W., Bradshaw, C.J.A. 2009, 'Causes and consequences of species extinctions', in Levin, S.A. (ed) *The Princeton Guide to Ecology*, Princeton University Press, Princeton, New Jersey.

BEATING CANCER

Australian scientists are discovering the tricks of one of humanity's deadly foes, writes Stephen Pincock

THIRTY-FIVE YEARS AGO, when Professor Ian Frazer was a young medical student in Scotland, doctors regularly decided not to tell patients they had cancer. Hiding the bitter truth was widely thought to be kinder. After all, there was often little that doctors could do to treat the disease, says Frazer, the 2006 Australian of the Year who helped develop a vaccine against cervical cancer. "It was considered better for them not to know."

Cancer treatment has changed radically since then. In the 1970s, when Frazer was in training, doctors might have hoped to cure 15-20 per cent of cancer patients by deploying the limited arsenal of surgical and radiation tools they had

at their disposal. Now, a transformation in cancer treatment means today's oncologists might expect to cure more than half of the cancer patients they treat.

"Now we have effective therapies that attack the fundamental problem in a whole range of common cancers, plus we have effective screening programs against some that we didn't have before," says Frazer, now the CEO and Director of Research at the Translational Research Institute in Brisbane. "If you did a 'then and now' analysis of cancer over the last 50 years it would be a very impressive shift."

Australian researchers have played a significant part in those successes, from the most fundamental laboratory research to the development of effective public health campaigns.

In spite of all these improvements, however, cancer remains the most common cause of death in Australia. Roughly half of us can expect to develop cancer by the time we are 80 and many common malignancies, such as pancreatic and ovarian cancers, are often deadly.

KNOW THINE ENEMY

David Bowtell, a cancer geneticist at Melbourne's Peter MacCallum Cancer Centre (Peter Mac), is optimistic science will make further strides against cancer in coming decades. He says that's because scientists have learned a lesson famously taught by a Chinese general in the sixth century BC.

"When I talk about cancer, I like to quote Sun Tzu. He said that if you know your enemies and know yourself, you need not fear a hundred battles," says Bowtell.

Bowtell explains that cancers develop in much the same way that computers crash. Where a bug in your laptop's operating system might cause it to fail, with cancer the errors in your 'operating system' are mutations in the DNA of a cell. These lead to cells growing and dividing in an uncontrolled way (see: **What is Cancer?**).

Over the past 20 years, scientists have developed DNA sequencing technology, allowing them to read the cancer's code and see how it's corrupted. This technology shows clearly that, even within the same tissue, a multitude of different mutations can lead to cancer.

Understanding the specific mutations that arise in an individual's cancer has already allowed scientists to design drugs targeting diseases such as breast cancer, some leukaemias and melanoma.

"We are already undergoing a paradigm shift away from traditional therapy in cancer to targeted therapy based on the mutations in an individual's cancer," says medical oncologist Ian Olver, chief executive of Cancer Council Australia.

Traditional chemotherapy and radiotherapy was based on a one-size-fits-all approach, causing severe side effects. Those traditional treatments target all rapidly dividing cells. So they kill healthy cells, such as hair follicle and gastrointestinal cells, as well as the cancerous cells. These side effects are why some cancer patients lose hair and experience nausea.

REVEALING THE DETAILS

The more scientists learn about the changes that drive individual cancers, the greater the number of treatment advances that will emerge, claims University of Queensland's Sean Grimmond.

In the near future, scientists will be able to decode the entire genetic blueprint of an individual's normal cells cost-effectively, then compare it to the complete series of errors in a tumour. "This is really a new tool for us to battle cancer," he says.

Australian researchers are participating in some of the large international projects that will make this a reality. Both Grimmond's group at the Queensland Centre for Medical Genomics and Bowtell's group at Peter Mac are taking part in the International Cancer Genome Consortium project, which is building an atlas to understand the major events that cause the 50 most common cancer types worldwide.

Australia's contribution on this project is to investigate pancreatic and ovarian cancers. So far, Bowtell's group has sequenced about 80 patients with ovarian cancer, looking for mutations associated with drug resistance. The Queensland group has analysed cancers from about 400 pancreatic cancer patients. According to Grimmond, their pilot work has more than doubled survival rates by helping doctors to decide on the most effective treatment for patients.

Beyond understanding changes in the genetic sequence underpinning cancer, researchers are also looking at other changes and modifications made to DNA, notes Susan Clark from Sydney's Garvan Institute.

"It's very clear that understanding the changes to the DNA code is not enough to understand how parts of the code are turned on and off in cancer cells," she says.

Scientists refer to the additional information that determines which genes are switched on and off as 'epigenetic' changes. Some of these changes are made in response to the environment, whereas other epigenetic 'marks' are made during early

In the near future, scientists will be able to decode the entire genetic blueprint of an individual's normal cells cost-effectively, then compare it to the complete series of errors in a tumour.

embryo development, shortly after the egg is fertilised. These include chemical modifications to the DNA, which alter how the DNA is packaged into each cell and how genes are switched on and off. Epigenetic marks in cancer cells can be substantially altered, leading to many changes in gene activity, Clark says.

Of course, this extra information will pose new challenges for doctors trying to interpret it to treat their patients.

"I think we're going to see a big emphasis on this kind of medicine across a whole lot of cancers, and that will really require a big change in the way we think about health care," says Grimmond.

FUTURE THERAPIES

In 2012, Australian scientists reported in the journal *Lancet Oncology* that some melanoma patients with advanced disease received a dramatic benefit when treated with a drug designed for use in people whose cancers had a specific mutation. The drug was able to shrink tumours that had spread to the brain, which is a common problem in melanoma, and added months to the lives of many patients.

"The findings are among the most important in the history of drug treatment for melanoma," claims study co-author Georgina Long from Melanoma Institute Australia.

The study powerfully illustrates the promise of future therapies. Yet it also highlights a major challenge: relapse. While many patients saw a major benefit, for most the response was limited to a few months. The cancer became resistant to the therapy. The drug was no longer effective.

"I think where the personalised medicine story is falling down at the moment, and where Australia has a grand challenge, is in treating patients who we aren't able to cure up-front and whose cancer comes back," says Bowtell.

In fact, future cancer treatment may involve managing cancer rather than curing it, adds Grimmond. Technology that allows scientists to study the sequence of a person's entire genetic make-up – their genome – might become a tool for routinely reviewing and fighting cancers as they change.

RISK RECKONING

Scientists know the genes some people inherit put them at increased risk of developing cancer. Perhaps the most infamous of these cancer predisposition genes are *BRCA1* and *BRCA2*, which can cause a small fraction of breast and ovarian cancers. In 2013, actress Angelina Jolie raised awareness of these genes when she announced that she'd had a double-mastectomy to reduce her breast cancer risk, after finding that she carried a high-risk version of the gene *BRCA1*.

Scientists such as Georgia Chenevix-Trench from the Queensland Institute of Medical Research are identifying other genetic variants that can increase vulnerability to cancer. She says: "That's something we've been focusing on hard for the past five or 10 years, particularly for breast and ovarian cancer."

The researchers have found close to 80 genetic variants that increase breast cancer risk and 20 for ovarian cancer. Although each individual variant only confers a very small increased

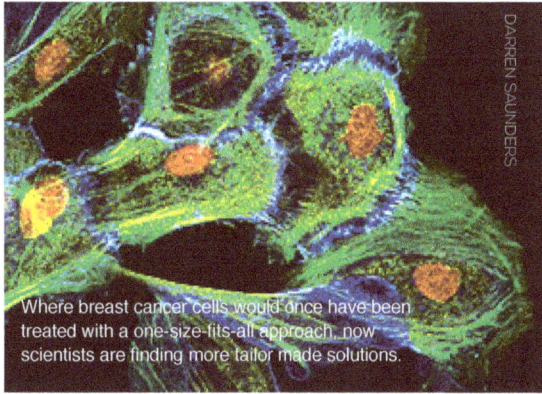

Where breast cancer cells would once have been treated with a one-size-fits-all approach, now scientists are finding more tailor made solutions.

cancer risk, in combination they may increase an individual's risk to the level of a *BRCA1* or *BRCA2* mutation.

Of course, even people who do not inherit cancer-risk mutations can develop cancer. Environment and lifestyle substantially increase the risk of developing multiple types of cancer. Such factors can combine with internal genetic mutations to disrupt the normal checks and balances on the growth and development of cells, leading to cancer. Smoking, sun exposure and asbestos exposure are known culprits.

Scientists know that one-third of all cancers are preventable with existing knowledge, says Frazer. Reducing smoking and alcohol consumption, avoiding obesity, staying out of the sun and ensuring people take part in screening programs and receive vaccines against hepatitis B and papillomavirus will protect millions from cancer.

If science can deliver those messages, along with new genomic treatments and better assessments of inherited risk, then cancer

WHAT IS CANCER?

Cancer is a word used to describe the many different diseases caused when abnormal cells divide without control.

Normally, cells grow and multiply in a regulated way, but errors in a cell's DNA, known as mutations, can cause that control to be lost.

Many cancers result from mutations in genes that regulate how often the cell divides, repair DNA mistakes, or tell cells when to die.

When cancers grow, they can disrupt essential body functions. For example, if the cancer grows in the liver or bones it can upset the body's delicate chemical balance. Or if it grows in the lungs, it can prevent enough oxygen from being absorbed.

Cancers can develop in solid tissue, leading to the formation of tumours, or in body fluids such as blood where they are called liquid tumours. Leukaemia is an example of a liquid tumour.

Some cancers are more aggressive than others. Aggressive cancers can spread through the body via a process called metastasis. This happens when cancer cells reach into blood or lymphatic vessels, allowing them to circulate through the body.

treatment will be unrecognisable to the doctors he trained with all those years ago, says Frazer.

Can those challenges be met? Frazer is "very optimistic".

STEPHEN PINCOCK is an author and science journalist based in Sydney.

FURTHER READING

Mukherjee, S. 2011, *The Emperor of All Maladies: A biography of cancer*, Scribner, New York City.

Cancer Council Australia is Australia's peak national non-government cancer control organisation, advising the Australian Government and other bodies on practices and policies to help prevent, detect and treat cancer.

The International Cancer Genome Project – based at the Wellcome Trust Sanger Institute – aims to identify sequence variants/mutations critical in the development of human cancers.

The World Health Organization Cancer Programme's mission is to promote national cancer control policies, plans and programmes, integrated to non-communicable diseases and other related problems.

Medical researchers are tackling the underlying causes and frightening symptoms of dementia, writes Stephen Macfarlane

MIND DECLINE

THE WORDS 'DEMENTIA' and 'Alzheimer's' are often used interchangeably. This is incorrect. Dementia is not a diagnosis in itself. Rather, it is a term used to describe any condition that leads to a progressive cognitive decline. It is estimated that dementia affects some 280 000 Australians. By 2030, this number is projected to double.

More than 100 causes of dementia are known, the most common being Alzheimer's disease, which accounts for between 60-70 per cent of all cases. Increasing age is the single biggest risk factor for Alzheimer's disease, and with modern medicine steadily increasing its capacity to extend the lifespan through the improved management of chronic disease, more people are surviving longer and to a very old age.

For those of us lucky enough to survive into our nineties, we will have a 50-50 chance of suffering Alzheimer's at that point.

Treatments for Alzheimer's disease are in their infancy. So-called cholinesterase inhibitors are the current mainstay of treatment. These drugs help damaged and dying brain cells – neurons – function better, producing a short-lived improvement in memory and cognition. They fail to modify the outcome of the disease in any way, however, and their benefit generally abates within two years of treatment.

There is a singular urgency, therefore, to develop treatments that target the pathology of the disease to produce a true disease-modifying effect rather than purely symptomatic benefits.

Two key pathologies are present in the brains

The challenge is to identify and treat people with the pathology before they develop symptoms of cognitive impairment.

of Alzheimer's disease sufferers. The first is the presence of abnormal Tau protein within affected neurons. When Tau is normally configured it allows transport of nutrients and brain chemicals called neurotransmitters along neuronal extensions. It provides the 'backbone' of these cellular transport channels. When Tau becomes involved in a chemical reaction known as hyperphosphorylation – regulated by a large molecule known as the GSK enzyme – it loses its structural rigidity and collapses in upon itself, forming a characteristic 'tangle'. The formation of a tangle, *per se*, is not causative for Alzheimer's disease, but represents the end stage of a chain of pathological processes involving Tau.

A protein called amyloid is the second disease-defining pathology in Alzheimer's. In its usual physiological form amyloid is ubiquitous in animal species. We all have the protein circulating within our systems, but when an abnormality occurs in its production, resulting in a molecule slightly longer than usual, amyloid can become involved in a series of chemical reactions that damage neurons.

Abnormal forms of amyloid attract various metals that are present as trace elements within our bodies. When these metals react with oxygen in the bloodstream they produce toxic products, known as free radicals, that damage cells and disrupt cell functions, leading to neuronal death.

While researchers differ over which of these two pathologies kickstarts Alzheimer's disease, they are following both lines of inquiry.

There are a number of medications known to target the abnormal phosphorylation of Tau. Lithium, for example, has been used for 60 years in psychiatry and is known to inhibit the action of GSK at low doses. High-dose selenium supplements may also help reverse

JOHN'S STORY

'John' was 63 when he lost his job as a senior manager in a multinational corporation. His colleagues thought he had lost his edge, or perhaps was drinking too much. His wife suspected that something was wrong, but was aware of his stress at work, and attributed any changes to that.

Six months later, John was diagnosed with Alzheimer's disease.

He began taking a cholinesterase inhibitor and shortly thereafter enrolled in a clinical trial of an antibody engineered to target the amyloid protein. He remains on the study drug and continues to be independent in many of his personal, domestic and community activities. After three years on medication he has declined at a rate some 30 per cent less than might have been expected in the absence of the study drug.

John is arguably one of the lucky ones, having been in a particular subgroup of patients who have been shown to benefit from this intervention. More research is needed to clarify the reasons for varying response rates in different patient groups within clinical trials.

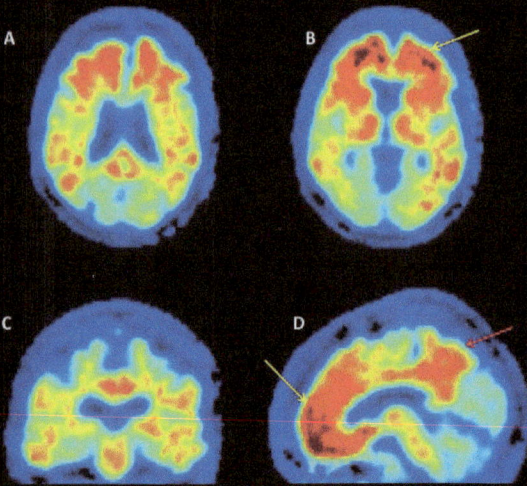

In contrast to the healthy brain on the left, the brain on the right shows regions in red (B and D, arrowed) with greater uptake of beta amyloid, protein pieces that clump together and gradually build up to form plaques in Alzheimer's-affected brains.

trials have demonstrated cognitive, or other clinical, benefits. In those that have, such benefits are largely confined to specific patient subgroups and to improvements on specific subscales of broader cognitive measures; their clinical significance is doubtful.

These disappointing results might best be explained by the stage at which people seek medical assistance, worried they might have Alzheimer's disease. While the oft-repeated claim that we only use 10 per cent of our brain power is an overstatement, it is certainly true that our brains hold more neurons than they need to work effectively. We have, consequently, a vast 'cognitive reserve'.

By the time somebody notices symptoms of memory loss, the capacity of our brains to compensate for the damage inflicted by Alzheimer's has, by definition, already been exceeded. A large amount of damage has already been inflicted and it may be unreasonable to expect improvement or even stabilisation of the disease, even at the point where difficulties first become apparent.

It's well known that amyloid pathology is present in the brains of those who subsequently develop Alzheimer's for at least 10-15 years before the earliest symptoms of the disease appear. But

Tau pathology by decreasing the amount of insoluble protein present. Melbourne company Velacor Therapeutics is at the forefront of this latter approach.

Illustrating the contributions from branches of science as diverse as immunology, biochemistry and genetics, several therapeutic approaches to the presence of abnormal amyloid are in the pipeline. They include various immunisation strategies, medications that inhibit the formation of amyloid by blocking the enzymes that lead to its production and others that promote dissolution of plaques by disrupting their toxic interaction with metals. Such a compound is PBT2, developed by Melbourne-based Prana Biotechnology, which has been in clinical trials for more than a decade now and is slowly progressing towards commercialisation.

While medications that target the pathology of Alzheimer's have been in clinical trials for more than a decade, these trials have generally had disappointing results. A number of studies have demonstrated that some drugs produce dramatic changes in measurable levels of markers of the disease in blood, but very few

AGE GROUP	INTERNATIONAL PREVALENCE (%)
60-64	1.3
65-69	2.2
70-74	3.8
75-79	6.5
80-84	11.6
85-89	20.1
90+	41.5

Table 1: The percentage of people within the specified age group who suffer from dementia

neurons continue dying throughout that time. So now the challenge is to identify and treat people with the pathology before they develop symptoms of cognitive impairment.

Alzheimer's cannot yet be prevented. Still, the development of radioactive tracer compounds that bind to amyloid and are detectable by brain scanners has opened the door to the possibility that asymptomatic individuals who harbour the pathology might be identified, leading the way to preventive treatments. These scans are being used in large-scale local research efforts, such as the Australian Imaging, Biomarker and Lifestyle Study, to help track the progress of both healthy volunteers and people with various stages of cognitive impairment in an effort to learn more about the progression and development of Alzheimer's pathology.

Prevention, in medicine, is always easier to achieve than cure. For those concerned about developing dementia in later life, there are a number of simple and powerful interventions that can help lower an individual's chances of developing the illness. The risk factors for Alzheimer's disease are now well known by clinicians, and are identical to those for cardiovascular disease and stroke. It is epidemiological research, rather than efforts focused within the laboratory, that has led to the clear elucidation of these risk factors over the past three decades.

Although old age and a family history are unmodifiable risk factors, other known contributors, such as high blood pressure, cholesterol, smoking and diabetes, can be reduced or controlled by lifestyle changes and medical intervention. It is the control of these variables on a national scale that gives modern medicine the greatest opportunity to impact the prevalence of dementia in the 21st century.

MANY KINDS OF DEMENTIA

While Alzheimer's is the most common form of dementia in Australia, other causes are also frequently diagnosed.

Vascular dementia accounts for 10-15 per cent of all dementia cases. Its onset is often relatively sudden and there can be periods of stability punctuating ongoing decline. A history of vascular disease is often present in other organs and evidence of multiple small strokes or of blood-vessel disease in the brain is often discovered during brain scans.

Lewy Body dementia – 10-15 per cent of all cases – is closely related to the dementia that can accompany Parkinson's disease. In its early stages, it is frequently misdiagnosed as Parkinson's, as patients tend to develop the motor slowing, shuffling gait and rigidity that characterise the better-known neurological condition. Other common symptoms include visual hallucinations, fluctuations in cognition, daytime sleepiness, vivid dreams, problems in the regulation of the autonomic nervous system – often experienced as urinary incontinence or falls – and extreme sensitivity to antipsychotic medications often used to treat hallucinations.

Frontotemporal dementia accounts for 5 per cent of cases and tends to strike younger people. For those in their fifties it is probably more common than Alzheimer's disease. Telltale symptoms include either behavioural and/or personality changes or early language difficulties, usually problems finding the correct words or a difficulty recognising objects for what they are.

ASSOCIATE PROFESSOR STEPHEN MACFARLANE is the director of Caulfield Aged Psychiatry Service in Melbourne, and associate professor of Aged Psychiatry at Monash University.

FURTHER READING

Alzheimer's Australia is the peak body providing support and advocacy for the more than 321 000 Australians living with dementia.

The Department of Health, 'What is dementia?', fact sheet, http://www.health.gov.au/dementia.

IT'S NOT ALL IN THE MIND

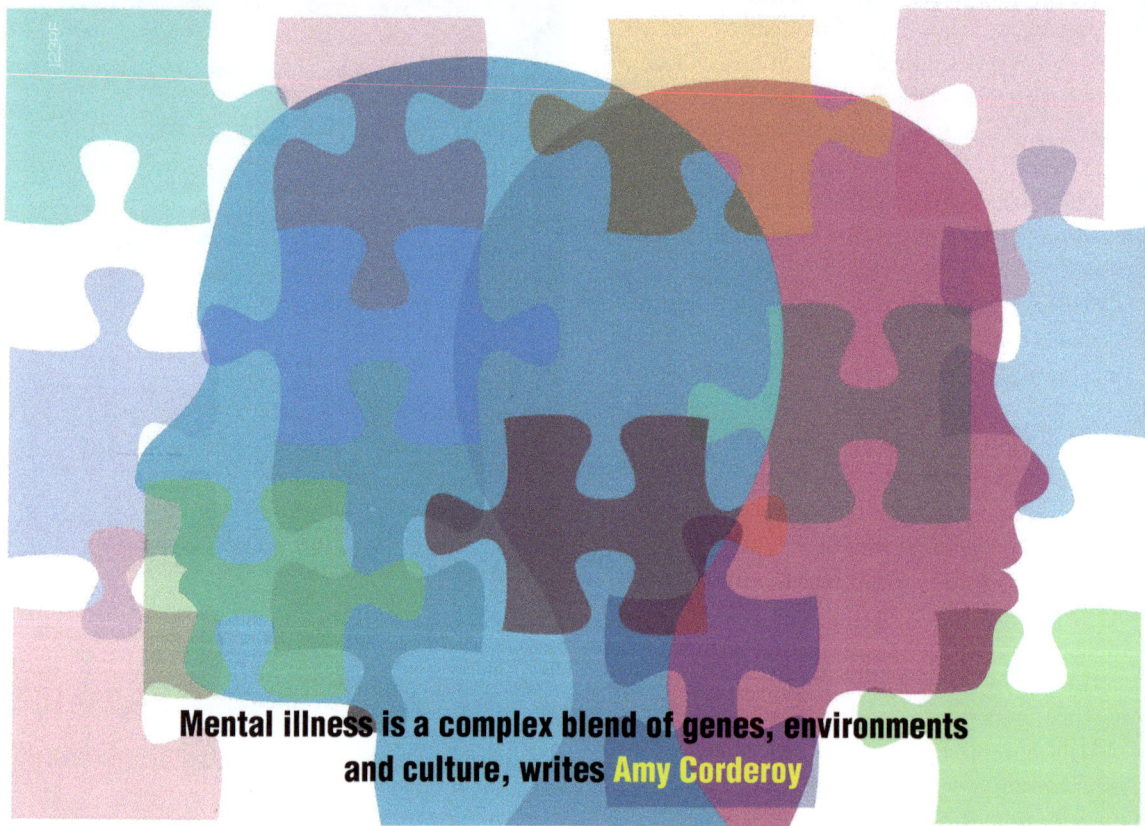

Mental illness is a complex blend of genes, environments and culture, writes Amy Corderoy

YOUR MIND IS racing. It's making connections between people, places, events, figuring out things you've never understood before.

"There is no such thing as a coincidence," says Matt (surname withheld). "Your mind is working things over and over."

Matt has schizophrenia and is describing his first psychotic episode. His brain made false connection after false connection. His heightened awareness focused on things that were not real.

"I just started to think everything I saw and heard was about me... I was terrified," he says.

Schizophrenia is a life-changing condition. Australians diagnosed with it die, on average, 25 years early. It is part of a spectrum of symptoms and experiences we call "mental illness", an enormously broad term that will apply to one in five Australians each year.

Understanding, treating, and preventing mental illness is one of the most complex and difficult tasks facing scientists. The research depth and breadth needed – from understanding the tiniest molecules to the broadest measures of population health – is enormous.

But research into mental illness holds the promise of spectacular gains: not only curing emotional pain but understanding how the most social organ in the body – the brain – is affected by the world around it.

Just what is a 'normal' behaviour is defined by culture. Our expectations of behaviour change over time and differ between societies. And as scientists look deeper into mental illness they are questioning the established 'objective' boundaries of disease.

For Matt, the difference between healthy and unhealthy is obvious. But sorting illness from normal emotional and personality differences is frustratingly difficult at times.

This tension between treating illness and 'medicalising' normal life came to a head in May 2013 with the launch of the fifth volume of the *American Psychiatric Association's Diagnostic and Statistical Manual of Mental Disorders* (DSM-V), which influences diagnosis, and shapes research, law and policy.

US psychiatrist Paul Chodoff summed up the difficulties in a 2005 letter to *Psychiatric News* proposing a new diagnosis: "the human condition". It would cover everything from distractibility, unhappiness, body image dissatisfaction and upset when things go wrong. The diagnosis "would facilitate insurance reimbursement… and encourage the quest for a drug to cure the disorder of being human," he wrote. Many critics of modern psychiatry argue Chodoff's joke wasn't far off the truth.

Among them is Allen Frances, architect of the fourth edition of the manual. "We were definitely modest, conservative and non-ambitious in our approach to DSM-IV," he says. "Yet we had three epidemics on our watch

FUTURE DRUGS

Medications for mental illness are known for their side effects and limited efficacy for some symptoms. Antipsychotics, for instance, increase the risk of weight gain, diabetes and elevated triglycerides and cholesterol.

But researchers like Karen Gregory hope future drugs – some already in early trials – will seek out new targets, with fewer side effects. A National Health and Medical Research Council (NHMRC) Overseas Biomedical Postdoctoral Research Fellow, Gregory says traditional antipsychotics target dopamine, which is involved in motor control, motivation, arousal, and reward.

"However, these therapeutics don't treat all the symptoms associated with schizophrenia," she says. "One of the key challenges is… how to target negative and cognitive symptoms, which are correlated with long-term patient outcomes."

Two promising neurotransmitters being targeted are glutamate, which has reduced signalling in schizophrenia, and acetylcholine, which modifies areas of the brain believed to have reduced functioning when someone has the condition.

Dr Gregory says new drugs targeting these transmitters could be available within 8-10 years.

– autism, attention deficit-hyperactivity disorder (ADHD) and child bipolar disorder."

When his taskforce reduced the number of symptoms for ADHD they thought it might increase by 15 per cent. Instead, diagnosis increased by 200 per cent.

In Canada, a 2012 study of nearly a million primary school children showed how subjective diagnosis can be. It found boys were 30 per cent more likely to be diagnosed with ADHD – and 41 per cent more likely to be medicated – if they were simply the youngest in their school year.

Critics fear the new manual will create a new generation of epidemics.

So controversial is it that, on the eve of its release, key treatment and research bodies have called for it to be abandoned. But they're split

Every thought we have changes the chemistry in our brains and that chemistry is defined by both environmental inputs and genetic inputs.

on whether the focus should shift to developing new, better, solely medical models of mental illness, or broader models that take into account both individual experience and measures of population health.

The US National Institute of Mental Health, the world's largest mental health research body, made waves by saying it would abandon the DSM in favour of a new biomedical model, which, it states, will "transform diagnosis by incorporating genetics, imaging, cognitive science".

It's an ambitious project. As yet, "no scientific studies to date have shown that a brain scan by itself can be used for diagnosing a mental illness or to learn about a person's risk for disease," the Institute says.

Sydney psychiatrist Tad Tietze quotes Samuel Beckett to describe the Institute's announcement. "Try again. Fail again. Fail better." He says such an approach will always be too narrow because it fails to pick up the underlying social causes of mental illness.

The British Psychological Society says attempts to pin down mental illness objectively on symptom lists, brain dysfunction or chemical imbalances are not evidence-based and are doing more harm than good.

"Patients can spend years rotating in and out of hospital without anyone sitting down and trying to help them make sense of their distress in terms that are personally meaningful to them," says Lucy Johnstone, a psychologist and member of the Division of Clinical Psychology with the society.

But what if these big picture social factors could be revealed under the microscope?

In his University of Newcastle lab, geneticist and molecular biologist Murray Cairns is working to discover what goes wrong in the brains of people like Matt. Cairns' work on microRNA has found these tiny molecules – which enable complex patterns of gene expression, or action – could be the key to understanding schizophrenia.

He says massive global genetic studies are pinpointing hundreds of genes potentially linked to the disorder. Cairns is not bound by the DSM. Instead, he seeks to understand gene expression in different subtypes of schizophrenia such as those with psychotic symptoms or cognitive impairment.

"In most people, schizophrenia is probably caused by a large number of (gene) variants that are actually common in the population," he says. The question is why some people with the variants develop the disorder while others don't.

MicroRNA could be the answer, says Cairns. It enables brains to change, responding to environments and experiences by forming new connections, turning genes on and off in the process.

Cairns thinks it's unlikely single microRNA molecules or genes will be identified that cause illness. Instead, he is looking at complex pathways, or combinations of changes, that may lead to similar types of symptoms.

"Everything boils down to molecules," Cairns says. "Healthy or unhealthy, every second there are changes going on. Every thought we have changes the chemistry in our brains and that chemistry is defined by both environmental inputs and genetic inputs."

Early-life poverty, stress, inflammation, maternal infection and circadian rhythm disruptions are among the many issues identified as potentially pushing people down a path towards illness.

And scientists are now discovering genetic risk factors thought to be linked to one condition are actually linked to multiple mental illnesses, with genetic changes initially investigated for schizophrenia now linked to conditions such as autism, ADHD and bipolar disorder.

In his 2010 book *Crazy Like Us*, Ethan Watters shows how socially and historically defined frameworks for expressing psychic pain can also influence illness progression.

He visits Zanzibar to explore the puzzling phenomenon that people with schizophrenia in developing countries seem to fare better than those in western countries – a reversal of the usual split of health outcomes between rich and poor.

There, Watters discovers that researchers have found that religious, fatalistic and non-individualist beliefs allow a more accepting attitude towards schizophrenia, decreasing emotional intensity between families. Lower levels of "high expressed emotion" are known to be protective against relapse.

Belief in commonly occurring spirit possession also helps explain psychosis. "The point was not that these practices were effective in combating the biological causes of schizophrenia," he writes. "Rather, they were simple examples of the sick person [being kept] within the social group."

President-elect of the Royal Australian and New Zealand College of Psychiatrists, Mal Hopwood, says the complicated story behind mental illness leaves the public feeling confused and perhaps bemused.

But in the end, "to consider either biological *or* social factors exclusively would really be missing the point," he says.

That's a complex web to untangle. The scientific examination of the spectrum of mental disorders – a spectrum as diverse as each individual who suffers from them – requires input from scientific disciplines as varied as anthropology and genetics. The lives and happiness of Matt and millions of others are at stake.

AMY CORDEROY is a journalist and the health editor of the *Sydney Morning Herald*.

FURTHER READING

Frances, A. 2013, *Saving Normal*, William Morrow, New York City.

Vaughan Bell's blog, *Mind Hacks*, http://mindhacks.com/.

Watters, E. 2010, *Crazy Like Us*, Free Press, New York City.

Metzl, J. 2010, *The Protest Psychosis*, Beacon Press, Boston.

National Institute of Mental Health Director's Blog, http://www.nimh.nih.gov/about/director/index.shtml.

Tad Tietze's blog, *Left Flank*, http://left-flank.org/.

Morrow, L. et al 2012, 'Influence of relative age on diagnosis and treatment of attention-deficit/hyperactivity disorder in children', *Canadian Medical Association Journal* 184(7): 755-762.

Corderoy, A. 2012, 'Another generation of epidemics', *The Age*.

PANDEMICS: LEARNING FROM THE PAST TO PROTECT THE FUTURE

Influenza remains the archetypal pandemic virus, with new strains arising periodically for more than a century, writes Dominic Dwyer

THE HISTORY OF humanity is replete with examples of disease scourges that have devastated human populations – infections such as Spanish flu, bubonic plague, smallpox, and tuberculosis. In the 14th century, bubonic plague – an infection of the lymphatic system resulting from the bite of an infected flea – killed an estimated 25 million people, or 30-60 per cent of the European population, while the 1918-20 outbreak of Spanish flu killed more people than all the military deaths in World War I.

Some of the old threats have now disappeared: for example, certain poliovirus strains have all but been eradicated and the smallpox virus was eradicated in the 1970s in a process overseen by one of the greats of Australian science, Professor Frank Fenner of Australian National University. However, the World Health Organization (WHO) still reports 1000 to 3000 cases of bubonic plague every year globally and new

viruses with pandemic potential have appeared in recent decades, such as HIV, Hepatitis C, and new influenza subtypes. Some new pathogens, such as Severe Acute Respiratory Syndrome (SARS) and the newly described Middle East Respiratory Syndrome Coronavirus (MERS-CoV) are significant, but fortunately outbreaks so far have been localised rather than becoming true global pandemics.

Of all the diseases that affect humans, it is the influenza virus that is most associated with pandemics, and it is the influenza virus that this essay is focused on.

INFLUENZA PANDEMICS THROUGH HISTORY

Although influenza pandemics have been described throughout history, differentiating influenza from other infectious diseases was difficult in the past. From 12th century Europe there were convincing descriptions of possible influenza pandemics, but determining whether

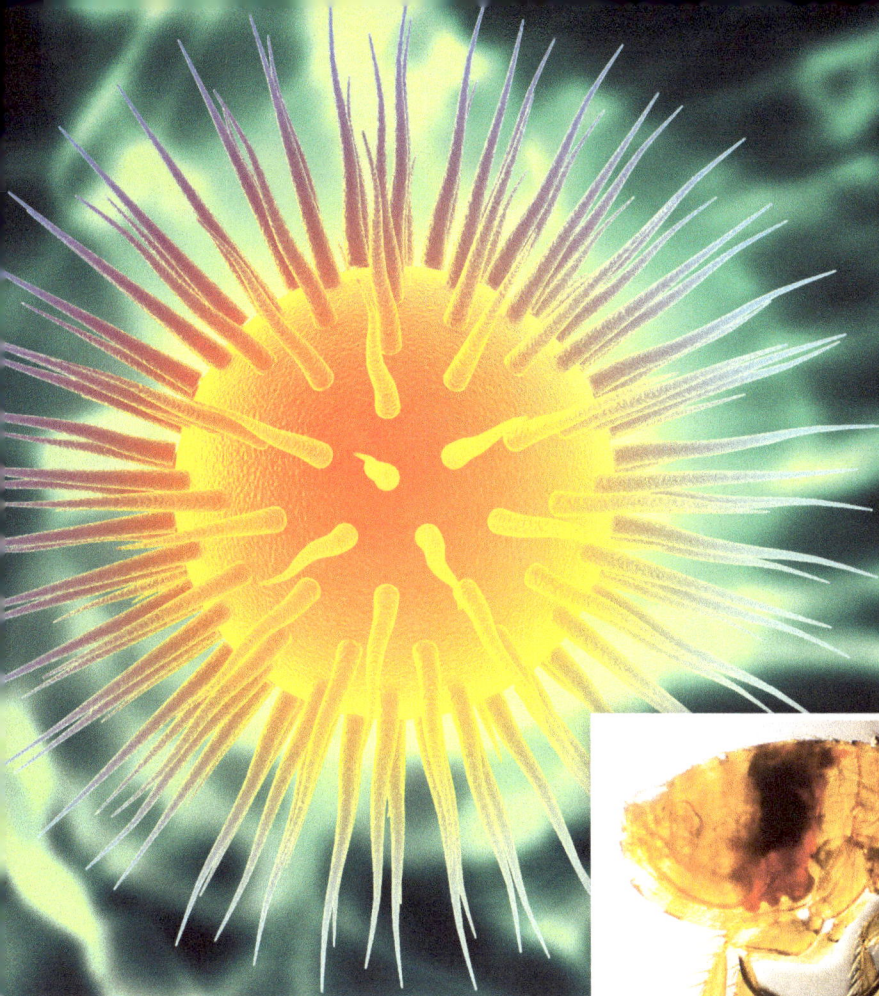

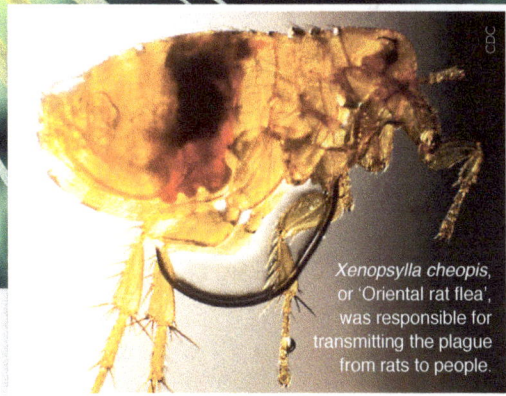

Xenopsylla cheopis, or 'Oriental rat flea', was responsible for transmitting the plague from rats to people.

these epidemics were truly global was impossible for many centuries (See: **Epidemic or pandemic**).

Perhaps the first modern description of an influenza pandemic occurred in 1889-1891: a pandemic that arose in Russia in the spring of 1889 and spread globally. This outbreak reached Australia and New Zealand in March 1890. In this pandemic, the secondary bacterial infections that cause pneumonia, and may complicate influenza, were first recognised. It was also observed that pandemic diseases may come in waves, sometimes becoming more severe with each surge.

The next great pandemic, often called Spanish flu, is the event to which all others have been compared. At least 50 million people died, and some 25 per cent of the world's population at the time was infected. Indigenous peoples, previously unexposed to the influenza virus, were the most severely affected, with up to 90 per cent of infected

people dying compared with one per cent of those in urban communities. This is likely because urban dwellers had a higher degree of pre-existing immunity to similar flu viruses that protected against the worst ravages of Spanish flu.

The origin of Spanish flu is still uncertain. It possibly started in China and then moved,

EPIDEMIC OR PANDEMIC

An epidemic occurs when the number of people infected is greater than expected within a particular country or region. A pandemic is different to an epidemic in that a pandemic infects many more people, crosses continents, and is usually due to a newly identified infectious pathogen. Pandemics are sometimes colloquially called plagues.

In the Walter Reed Hospital, Washington D.C., a flu ward was set up in an open gallery to accommodate patients during the Spanish flu pandemic.

through immigration, to the United States. The influence of World War I was strong in the distribution of this pandemic; it spread through the Western Front in Europe, and then quickly scattered around the world as soldiers returned home after the war. It took around 10 months for the virus to reach Australia, most likely due to the length of time it took for ships to reach here, and the strong quarantine measures initially imposed on arriving ships.

Once the virus arrived in Australia and the Pacific, the impact was significant; particularly in some of the remote Pacific islands where mortality rates as high as 25 per cent were reported. Clinical descriptions at the time included the sudden onset of what would now be diagnosed as acute viral pneumonitis (inflammation of the lungs), with rapid progression and death. Secondary bacterial infections such as pneumonia were also common, and may well have been the major cause of death.

An important feature of the 1918-20 pandemic was that deaths occurred mainly in young adults, in contrast to what is usually observed in the very young and elderly during the annual winter influenza epidemics.

SUBSEQUENT FLU PANDEMICS

There were further pandemics during 1957-58, 1968-70, 1977-81 and 2009, caused by different influenza A subtypes. Normally the new pandemic subtype displaces previous circulating influenza subtypes; however, sometimes subtypes can co-circulate, which happened in 1977-81 (A/H1 with the previously circulating A/H3) and 2009 (a reassortant A/H1N1 with the older A/H3N2) (See: **Reassortant influenza viruses**).

The 2009 influenza A/H1N1 pandemic spread quickly – in contrast to earlier pandemics, air travel was very important in its rapid global movement. As with Spanish flu, younger adults were most frequently affected. Very elderly individuals, alive in 1918 or the decade following, were less affected as they had A/H1N1 immunological memory through antibodies formed during childhood exposure to the Spanish flu subtype.

INFLUENZA ISOLATION AND IDENTIFICATION

Although influenza pandemics have been described throughout history, differentiating influenza from other infectious diseases was difficult in the past.

Spanish flu 1918-1920

"I had a little bird and its name was Enza, I opened the window and in-flu-enza"

Children's jump-rope rhyme heard nationwide during the height of the pandemic.

The influenza virus was first isolated from pigs and then grown in laboratories using chicken eggs, and then later using new methods of growing viruses in laboratory cell cultures. These methods were enhanced by Sir Macfarlane Burnet and others in Australia, and are still used today in specialised virology laboratories.

The isolation of the influenza virus from a human occurred two years later in 1933 and accompanying epidemiologic, laboratory and public health investigations now allow the accurate assessment of influenza pandemics. Today, we are able to grow and culture the influenza virus in the laboratory and use molecular biology techniques to determine which virus subtype is responsible for the infection.

In the 1980s, Peter Colman's protein crystallography group at the CSIRO in Melbourne discovered the crystal structure of the influenza neuraminidase protein, found on the surface of the virus, which is critical for the influenza virus lifecycle (See: **What is influenza?**). This discovery fast-tracked the development of the now widely used neuraminidase-inhibitor drugs, such as Relenza and Tamiflu.

Advances in science also now allow us to determine the mechanisms by which these viruses attach to receptors in birds, pigs or humans; their sensitivity to antiviral drugs; their likelihood of responding to vaccination; and to understand their spread in the community, including closed communities such as university colleges, schools, and aged care facilities.

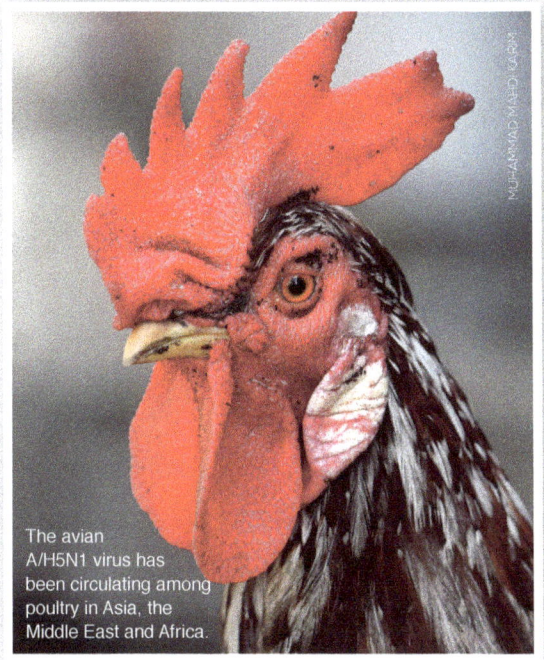

The avian A/H5N1 virus has been circulating among poultry in Asia, the Middle East and Africa.

PREVENTING FUTURE PANDEMICS

Science has helped us come a long way since Spanish flu. As a result of work by the CSIRO and others, antiviral agents were available for the A/H1N1 pandemic in 2009, although their value was not always easy to determine. An influenza A/H1N1 vaccine was available soon after the first wave, meaning that where it was routinely available, the subsequent waves – and there have been at least three – were probably diminished. Unlike many other countries, Australia has influenza vaccine manufacturing capacity.

Advances in medical technology – such as the antibiotics used to treat secondary infections following infection with the flu virus, and

REASSORTANT INFLUENZA VIRUSES

For a new influenza subtype to cause a pandemic, it must not only move into humans from its animal source, but must also be readily transmissible from person to person. Reassortant influenza viruses occur when influenza genes from different animal sources mix to become a novel subtype. For example, the 2009 pandemic virus contains genes from influenza viruses from humans, birds and pigs from both North America and Asia, resulting in a novel virus that then spread worldwide.

The surface of influenza A viruses are covered in protein spikes – haemagglutinin and neuraminidase.

WHAT IS INFLUENZA?

Influenza is a virus characterised by its high genetic variability and can be classified into three types – A, B and C. Influenza A is the most worrisome and virulent type for humans. Influenza C viruses typically cause only mild infection.

Influenza A viruses are classified on the basis of their genes which lead to the production of protein spikes – haemagglutinin and neuraminidase – that cover the surface of the virus. Haemagglutinin has 18 subtypes and neuraminidase has 11 subtypes. These subtypes are denoted by numbers, such as in H5N1.

intensive care medicine – probably contributed to a significantly reduced death rate in 2009 compared with 1918-20. By the same token, immune suppression in the general population – due to therapy for other illnesses, diabetes and respiratory disease, transplantation and other immune-suppressive diseases and newer problems such as obesity – are likely to have

contributed to the mortality and morbidity attributable to the influenza outbreak.

Other influenza viruses continue to pose concerns for future pandemics. The avian A/H5N1 virus has been circulating in poultry throughout various countries in South East and Central Asia, and the Middle East and Africa, for some years. Although human infections are associated with a very high mortality, significant person-to-person transmission has not occurred. Thus, although the potential remains, a pandemic has yet to develop.

Similarly, the emergence of A/H7N9 influenza in China in the last 12 months has raised concerns about worldwide spread. Fortunately, this has not happened and cases remain essentially limited to China. However, these examples show that although new influenza viruses continue to emerge, we cannot predict the likelihood of them causing a pandemic.

New scientific technologies have allowed an understanding of how these viruses

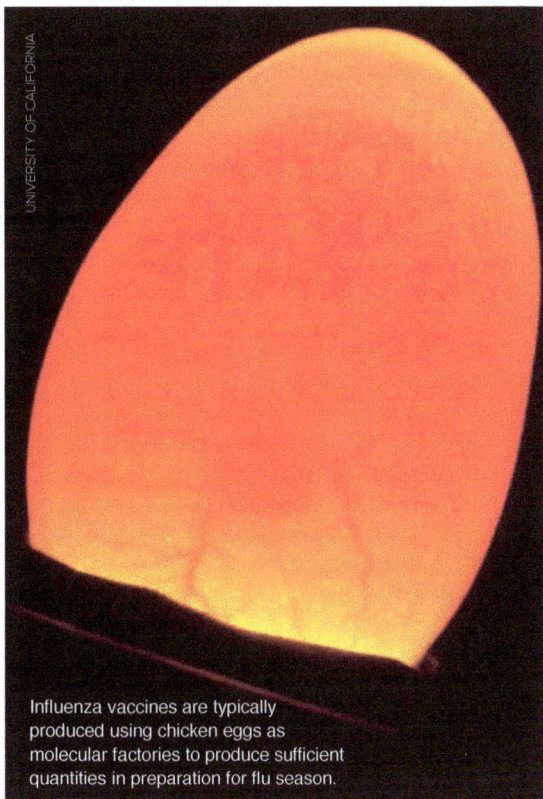

Influenza vaccines are typically produced using chicken eggs as molecular factories to produce sufficient quantities in preparation for flu season.

Melbourne (VIDRL) and Perth (PathWest). The WHO Collaborating Centre for Influenza Research and Surveillance in Melbourne is one of five collaborating centres worldwide, and has a particular role in determining the annual influenza vaccine components.

THE FUTURE OF PANDEMICS

The ease, volume and rapidity of air travel, the increasing world population, diminishing vaccination rates, changes in animal husbandry (especially of birds and pigs) and the increasing frequency of mass gatherings (sporting, cultural and other such events) all contribute to rapid spread of new viruses. These features now guide approaches to public health management and preparedness for viral infections. Surveillance, both in the clinic with infected patients and in the laboratory with virus strains, underpins pandemic responses, at state, national and international levels.

The contribution of scientific advances to allow rapid identification and isolation of influenza virus strains together with these advanced surveillance techniques means that we are more prepared than we have ever been to deal with the pandemic threats of the future.

cause disease: for example, the ability to detect the 1918-20 virus in old clinical specimens using molecular techniques, and the ability to 'recreate' these viruses in the laboratory. Despite careful attention to biosecurity, this has raised ethical concerns on how such research should be undertaken; work is currently being done to address these concerns.

Rapid and sensitive laboratory tests to detect new viruses are a focus of research and development in Australia, especially at the WHO National Influenza Centres in Sydney (ICPMR at Westmead Hospital),

PROFESSOR DOMINIC DWYER is a medical virologist and infectious diseases physician in the Centre for Infectious Diseases and Microbiology Laboratory Services, located in the ICPMR at Westmead Hospital, Sydney. He has a clinical and research interest in viral diseases of public health importance.

FURTHER READING

Nicholson, K.G., Webster, R.G., Hay A.J. 1998, *Textbook of Influenza*, Blackwell Science, Oxford.

WHO Collaborating Centre for Influenza Research and Surveillance (Melbourne) – part of the WHO Global Influenza Surveillance and Response System established in 1947 to monitor the frequent changes in influenza viruses.

Centers for Disease Control (USA), 'Seasonal influenza: Flu basics', http://www.cdc.gov/flu/about/disease/.

Doherty, P. 2012, *Sentinel Chickens: What Birds Tell Us About Our Health And Our World*, Melbourne University Press.

POPULATION HEALTH: UNDERSTANDING WHY DISEASE RATES CHANGE OVER TIME

Research on sets of individuals, whether in the community or laboratory, only partly explains the causes of disease, writes **Tony McMichael**

THE STREAM OF media stories about the seemingly ever-increasing numbers of Australians who are overweight or, frankly, obese continues relentlessly. Meanwhile, other media reports tell us excitedly about every newly discovered gene in laboratory rats that affects weight gain, about obesity that runs in families – "it must be genetic" – or about overweight people being more likely to be sugar and fat addicts.

So you might conclude from this that the overall problem is due to individuals' abnormalities of genes, metabolism or behaviour. Well, that conclusion is largely wrong.

The mistake is to have looked only *within* the population and noticed some distinctive things about the most overweight people. The health research arena offers us many such individual-level observations: individuals who have a stroke from a burst blood vessel in the brain often have very high blood pressure; individuals who smoke are much more likely to have lung cancer; individuals experiencing heart attack typically have personal histories of eating a lot of fatty foods. These are important initial findings, but they fail to consider the larger-framed and more interesting question about trends in the health profile of the whole population.

THE POPULATION PERSPECTIVE

A different perspective emerges, however, when these health problems are viewed on a larger canvas, at population level. Focusing first on stroke, national statistics show that the average blood pressure in the adult Australian population drifted upwards during the 1920s to 1940s and then in the late 1950s began to come down.

When I studied medicine in the 1960s the 'normal' blood pressure for a person was said to be 100 + Age. So a 60-year-old was expected, on average, to have a top (systolic) pressure of 160. That systolic blood pressure of 160 in 60-year-olds, then considered 'normal', would today be deemed significant hypertension, needing drug therapy. Meanwhile, the annual *rate* of death from stroke in Australia followed a similar up-then-down time-trend.

That time-trend raised an important question. Why did average blood pressure in the Australian population-at-large rise during the first half of last century and then fall over the final four decades? Further research indicated that the rise was related to the increase over time in the amount of available food per person in Australia, gradual increases in average body weight, and in the intake of salt (including as a food preservative in the pre-refrigeration era). The subsequent post-1950s downturn was largely attributable to the advent of new blood pressure-lowering drugs and a decreased reliance on salted foods.

Similar population-level explanations exist for the other examples. As the Australian diet became less fat-laden during the latter 20th century, partly influenced by the introduction of Mediterranean dietary habits and partly due to increases in public understanding, rates of heart attack decreased (clinical interventions, too, were becoming more effective). The adding of fluoride to reticulated drinking water (except for Queensland!) was followed by a reduction in childhood dental decay.

That larger-canvas perspective is the essence, the challenge, of *population* health. On that view, the population itself, as a living super-organism, has its own ecology – its own way of living, working, eating, relaxing and socially interacting. Changes in human ecology over time make big differences to patterns of poor health, diseases and premature death. So, our society will continue to miss the key point in much public discussion and policy-making if the primary focus on individual-level factors persists, buoyed by the prevailing philosophy of neoliberalism and individual responsibility. Good population health research provides guidelines for society as to what should be modified in our shared way of living; it addresses 'Big Policy' options.

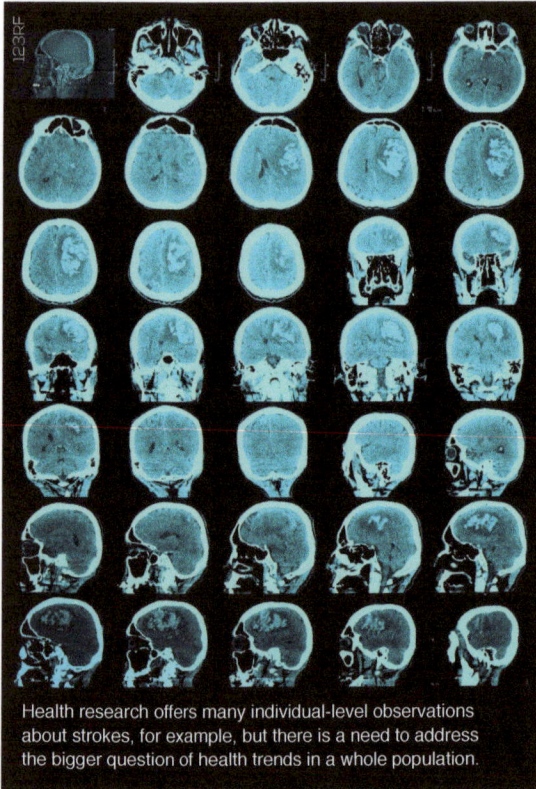

Health research offers many individual-level observations about strokes, for example, but there is a need to address the bigger question of health trends in a whole population.

SEEKING EXPLANATIONS FOR THE RISE OF OVERWEIGHT AND OBESITY

Now, returning to the overweight-obesity topic, undoubtedly some individuals are at higher risk than others. Populations always comprise a range of individuals with minor, genetically-determined, differences in their metabolism, including how food energy is either stored or burnt. Indeed, that's generally true for most diseases.

When a population lives in a way that generally fits natural human biological needs, then those minor differences rarely matter. But when the population's rate of some particular health disorder rises over several decades, that means something more general is amiss.

The recent rapid rise in overweight-obesity can't be due to good or bad genes; their frequency within the population is essentially unchanging. It can't be because the proportion of aberrant personalities and behaviours has increased; those are fairly culture-bound and don't change

A larger-canvas perspective is the essence, the challenge, of *population* health. On that view, the population itself, as a living super-organism, has its own ecology…

en masse. There must, then, be some change in the physical and social environment in which the population is living, a change in human ecology.

The basic problem is that, over the past 4-5 decades, the entire population has undergone a shift in average personal daily 'energy budget' – the ratio of energy intake (from food) compared to the amount of energy expended (mobility, working, recreation, and household activities). Hence, the population-level domain offers the best opportunity for substantial and enduring solutions – unless we use mass medication (analogous to fluoride in drinking water) to suppress appetites, or prohibit people from sitting down for more than four hours per day!

This population health perspective on overweight-obesity leads us into more creative, community-wide and far-reaching solutions. These include improving urban and suburban design to facilitate walking and cycling; upgrades in urban transport to reduce reliance on cars (good for urban air quality too); better recreational facilities and more green space for physical activity; and changes in work habits so that we spend less time glued to desks and computer screens without moving more than a few finger muscles for 2-3 hours at a time.

OTHER IMPORTANT ROLES AND INSIGHTS FROM THE POPULATION APPROACH

The population health perspective has other important roles. Two examples are illustrative. First, this perspective prompts comparison of health risks, experiences and needs of sub-populations: suburbs of lower and higher socio-economic position, males and females, different age-groups, ethnic groups and urban versus rural. And that provides an evidence base for much social policy.

INFLUENCES ON THE OCCURRENCE OF MULTIPLE SCLEROSIS AT POPULATION, INDIVIDUAL AND MOLECULAR LEVELS: THE FUN OF PIECING TOGETHER A JIGSAW PUZZLE

Multiple sclerosis (MS) is an autoimmune disease in which the body's immune system inadvertently attacks and destroys the insulating protein lining of the central nervous system, the myelin. In the ongoing Australian 'Ausimmune' study of risk factors for MS, the rate of occurrence of the preclinical MS condition (the 'first demyelinating event') ranges from 2.1 per 100 000 people per year in Brisbane to 8.7 per 100 000 in Tasmania. A fourfold increase as you travel south. There is no evidence of greater genetic susceptibility in the Tasmania population compared to Brisbane, which suggests that the observed difference in rates is due to some locally pervasive environmental factor. This population-level observation, similar latitude gradients in Europe and several other countries, suggests that the lower the local population's level of sun (solar ultraviolet) exposure, the higher the MS risk.

Ultraviolet radiation is known to affect the body's immune system. One pathway may involve vitamin D, largely produced within the skin by sunlight exposure. Average vitamin D levels are higher in Brisbane than Tasmania – indeed actual vitamin D deficiency is more common in Tasmania. Overall, we're finding that both lower levels of sun exposure and lower vitamin D levels are linked to an increased risk of having a first demyelinating event.

Genetic studies have shown that variants of several genes affect the body's response to vitamin D, and some may affect MS susceptibility more directly. So, individual genetic make-up may influence which individuals are most likely to develop MS. We and others have also considered epigenetic influences, whereby the lifelong activity of certain genes is, in effect, switched on or off in very early life by contact with exogenous factors, particularly dietary components. Some dietary molecular fragments can attach permanently to a particular gene during foetal life and switch the gene on or off.

Epigenetic phenomena contribute to a variety of disease processes. This, interestingly, is a case of nature *via* nurture.

Second, it helps us understand and control the spread of infectious diseases. Occasional cases of measles occur anyway, but when an epidemic breaks out it indicates that the population's 'herd immunity' has been lost. That is, an insufficient proportion of oncoming youngsters have been vaccinated, such that the measles virus can now maintain active circulation within the population.

A further role is that observations at population level can provide clues leading to more detailed studies. Our Environment and Health research group at the Australian National University's (ANU) National Centre for Population Health and Epidemiology working with a network of collaborators, has discovered one such clue by comparing regional rates of the autoimmune disease multiple sclerosis (MS) especially in reference to latitude and sun exposure. (See: **Influences on the occurrence of Multiple Sclerosis at population, individual and molecular levels: the fun of piecing together a jigsaw puzzle**).

THE KEY MESSAGE: UNDERCURRENTS CAN PROVIDE GREATER INSIGHTS THAN SURFACE RIPPLES

To reiterate, the key message here is that understanding rising rates of diseases in populations often requires attention to the underlying 'upstream' population-level influences. If a whole population eats a high-fat diet, then the annual rate of heart attack will be higher for that population than in others. Yet if, for example, every individual within that population eats very similar amounts of fat, but 30 per cent of them smoke and 70 per cent do not, then individual-level studies will mistakenly indicate that smoking, not dietary fat, is the main cause of heart attacks. Policy priorities may then be misdirected.

The obesity problem stems from the population's shift in average personal daily 'energy budget' – input outweighing output.

Similarly, to reduce MS rates we have learnt that a population should maintain a sufficient but not excessive exposure to sunlight, and avoid vitamin D insufficiency. As both work and recreation are tending to move indoors in modern Australia, average personal levels of sun exposure within the population are declining. At the individual level, there are some opportunities for family counselling based on genetic testing, although the role of genes in MS appears modest.

So the lesson is clear. Bifocal lenses should be used when examining the causes of disease, considering influences at both population and individual levels. Population health research often provides the best basis for the long-term lessening of various diseases in the population. With overweight and obesity, the real challenge, and long-term pay-off, is at the whole-community, or population, level – getting our planning of living environments and facilities and our way of daily life back in kilter with natural human biological needs, including restoring an input-output energy balance.

And that is where the challenge, the satisfaction and the fun of thinking in population health, resides – in thinking more broadly, and integrating one's ideas and strategies with those of interesting colleagues from other disciplines and sectors.

PROFESSOR TONY McMICHAEL is a medical graduate and epidemiologist at the Australian National University. During 2001-2012 he headed a research program on the health risks of climate and environmental change, present and future. Associate Professor Robyn Lucas assisted with the text on p55.

FURTHER INFORMATION

McMichael, A.J. 2013, 'Australia's Health: Integrator and Criterion of Environmental and Social Conditions', chapter in: Raupach M., McMichael A.J., et al (eds): *An Environmentally Sustainable and Socially Equitable Australia, by 2050*, Australian Academy of Science Press.

Lynch, J., Smith, G.D., Harper, S. et al 2004, 'Is Income Inequality a Determinant of Population Health? Part 1, A Systematic Review', *The Milbank Quarterly* 82(1): 5-99.

SCIENCE FOR OUR DAILY BREAD

The remarkable development of the Australian wheat industry clearly illustrates the role science plays in securing our daily bread, writes Snow Barlow

AGRICULTURE BEGAN with the domestication of wheat and barley in ancient Mesopotamia (part of the Middle East based on modern Iraq and Syria) more than 10 000 years ago. The food stability provided by these temperate cereal grains that could be stored over winter, enabled the development of 'settled' societies and allowed people free time for intellectual pursuits, culminating in the science we know now. It is on the modern version of this science that our civilisation will now depend in our efforts to ensure the world can feed the nine billion people projected by 2050.

In 1798, English scholar Thomas Robert Malthus predicted that populations would soon outpace food production: "The great law of necessity which prevents population from increasing in any country beyond the food which it can either produce or acquire, is a law so open to our view ... that we cannot for a moment doubt it."

Despite these dire predictions, over the past 150 years food production has generally kept pace with the seven-fold human population growth, from about one billion to today's seven billion plus. Science has played a pivotal role in increasing crop yields by up to tenfold over those of our ancient ancestors. Today, wheat, rice and maize (corn) supply more than 60 per cent of the globe's carbohydrate demand, essentially humanity's daily fuel.

Despite dire predictions, over the past 150 years food production has generally kept pace with the seven-fold human population growth.

The rise of Australia's wheat industry to one of the major global wheat exporters – with the United States and France – illustrates the role science plays in securing our daily bread. Globally, wheat accounts for one-third of total cereal production, or 700 million tonnes per year. It is grown across a broad range of temperate environments, including the drier edges of the world's great cropping areas.

The Australian wheat belt is a good example, stretching some 4000 km around the drier edges of the cropping zone, from Emerald in central Queensland to Geraldton in central Western Australia. In this region, 38 000 wheat farmers produce 38 million tonnes per year valued at AUS $8.5 billion – 5 per cent of world wheat production and 14 per cent of world wheat trade.

CONSERVATION AGRICULTURE

Conservation agriculture is a farming system developed in recent decades to conserve rainfall in soils, resulting in significant wheat yield increases. Moisture loss from soil has been reduced by replacing traditional mechanical cultivation of land – to control thirsty weeds during the summer – by chemical weed control that requires minimal tillage or soil disturbance. Additionally, the wheat stubble that remains after harvesting is left in place instead of being removed by burning or cultivation, again conserving water in the soil. Weeds are controlled by herbicides and crops are directly planted into the previous year's stubble.

Global Positioning Systems (GPS) have been an essential part of the development of these new farming systems. GPS allows farmers to sow seeds at an accuracy of 20 mm, directly between adjoining stubble growths from the

Figure 1: Australian wheat yields from 1860 to 2012

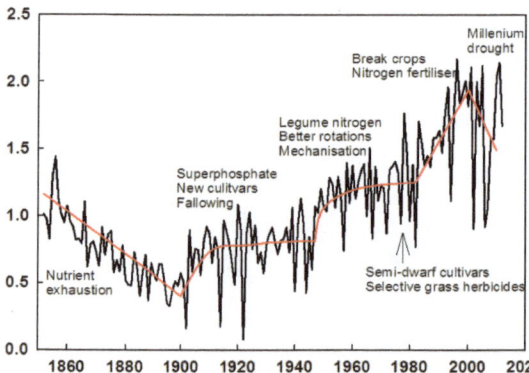

The black line is annual yield and the red lines are averages for each decade. Explanations for the trends are from Donald (1965) and Angus (2001) updated to 2012. (Graph kindly provided by John Angus, CSIRO).

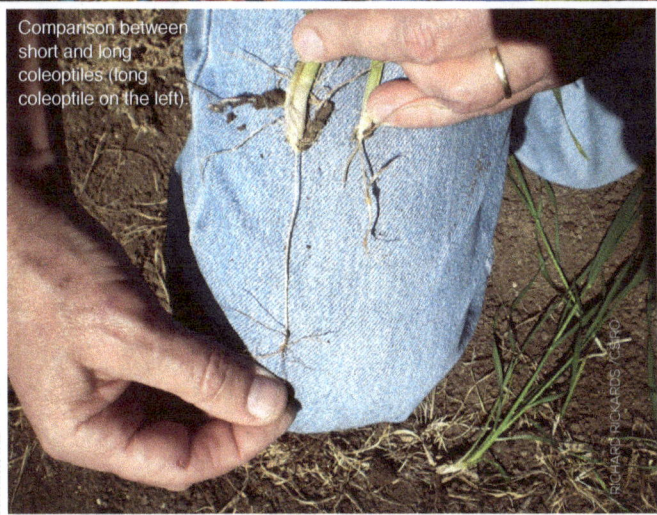

Comparison between short and long coleoptiles (long coleoptile on the left).

previous year, to protect further the emerging crop from the weather (see **Figure 1: Australian wheat yields from 1860 to 2012**). Because of this technology farmers don't drive around and around their paddocks. They drive up and down, reducing 'overlap' and using 5-10 per cent less time and fuel. This precise and repeatable positioning allows the farmers to set up 'tram tracks' for farm machinery in their paddocks, thereby minimising the compacted area. Reduced soil compaction can provide a further 5-10 per cent increase in yields.

Scientific advances in plant disease control have also contributed to increasing wheat yields. Wheat fields are complex biological systems where disease organisms, particularly those causing root diseases, can multiply if the same crop type is grown each year. As these disease organisms are usually specific to each crop, the build-up of disease organisms and resulting yield decreases can be averted by growing a different crop type – a so-called 'break crop' such as canola or chickpeas – as part of a crop rotation that can increase the yield of the subsequent wheat crop by as much as 20 per cent.

DEVELOPING NEW WHEAT VARIETIES

To reduce further the problem of getting the crop sown before winter, crop physiologists are seeking to exploit genetic variation in 'coleoptile' length. The coleoptile is the strong tube that emerges from a germinating wheat seed and grows towards the soil surface while protecting the soft first leaf within it. Current wheat varieties have coleoptiles some 50 mm long, effectively limiting their sowing depth

For every carbon dioxide molecule it absorbs, a plant loses 100-1000 molecules of water.

within the leaf evaporates and is lost from the leaf to the surrounding air (see: **A brief guide to photosynthesis**).

For every molecule of CO_2 absorbed by the leaf in photosynthesis between 100 and 1000 molecules of water are lost to the atmosphere through evaporation. This is why plants need so much water.

In the 1970s, CSIRO agricultural scientist John Passioura reasoned that Australian wheat yields are limited by the quantity of water available to the crop as the wheat grain grows to maturity and by the crop's efficient use of water. The challenge was how to measure plant water-use efficiency in a practical way.

At about that time, Graham Farquhar of the Australian National University and his American colleagues discovered that atmospheric CO_2 incorporating two isotopes or types of carbon, C^{13}/C^{12}, diffuses into the plant leaf at different rates because of the differing atomic weights of the carbon isotopes. These different rates are reflected in the composition of the plant. Farquhar and colleagues then showed that the ratio of C^{13}/C^{12} could be used in a practical way to measure the water-use efficiency of the plant as the ratio is a measure of the rate of CO_2 absorption through the leaf stomata.

Farquhar teamed up with Passioura's colleague, CSIRO wheat-breeder Richard Richards, to demonstrate that the C^{13}/C^{12} ratio of different varieties of wheat could be used to measure their water-use efficiency. Richards has used this technique to identify and breed a new generation of more water-efficient wheat for the Australian environment.

to 50 mm. However, varieties do exist with coleoptiles as long as 150 mm (see picture, p59). Using wheat varieties with this length would allow seed to be sown into the subsoil where moisture from summer rains is available, enabling crop establishment at the correct time and likely resulting in increased yields.

Land plants, including crops, use water as a means of transporting nutrients from the soil to their leaves where photosynthesis occurs. During photosynthesis, as plants open the small holes (stomata) in their leaves to absorb carbon dioxide (CO_2) from the atmosphere, the water

A BRIEF GUIDE TO PHOTOSYNTHESIS

Photosynthesis is the basis for nearly all life on Earth as it captures energy from the sun's rays (light), combines it with CO_2 to produce carbohydrates – the primary energy source. In the process, oxygen is released which provides the oxygen we breathe from the atmosphere.

$$\text{Carbon Dioxide (CO}_2\text{) + Water (H}_2\text{O)} \xrightarrow{\text{Light energy}} \text{Sugar (C}_6\text{H}_{12}\text{O}_6\text{) + Oxygen (O}_2\text{)}$$

Australian wheat yields have increased by 50 per cent in the last 25 years.

DRY SOWING

Australia's highly variable rainfall is becoming even more variable in the wheat belt due to climate change. This can be challenging for farmers who have traditionally relied on autumn rains to sow their crops before winter. Often summer rains have resulted in good moisture in the subsoil, but the surface soils where the seeds are sown are dry. Enterprising farmers have adapted to this variability and now sow their seeds into the dry surface soils, so-called 'dry sowing', so that they are ready to germinate when the rains come. This practice combined with the new long-coleoptile varieties described previously provides Australian farmers with a way to adapt to the changing rainfall patterns resulting from climate change.

CONCLUSION

The additive effects of these scientific advances on potential wheat yields is clearly shown in the graph on page 54, illustrating why Australian wheat yields have increased by 50 per cent in the past 25 years. These figures give strong encouragement that the continued application of science to Australia's most important food crop will lead to further yield increases into the future. As the global population increases beyond nine billion, applying good science to our food systems is the only way they will all receive their daily bread.

PROFESSOR SNOW BARLOW is a plant physiologist and agricultural scientist. His research encompasses plant water-use efficiency, viticulture and impacts of climate change on agriculture, water management and global food security. He is currently Foundation Professor of Horticulture and Viticulture at the University of Melbourne.

FURTHER READING

Ray, D. 2013, 'Crop crisis: Why global grain demand will outstrip supply', *The Conversation*, http://theconversation.com/crop-crisis-why-global-grain-demand-will-outstrip-supply-15298.

Prime Minister's Science, Engineering and Innovation Council 2010, 'Australia and food security in a changing world', PMSEIC, Canberra.

Linehan, V. et al 2012, 'Food demand to 2050: Opportunities for Australian agriculture, Research by the Australian Bureau of Agricultural and Resource Economics and Sciences'. Paper presented at the 42nd ABARES Outlook conference 6-7 March 2012, Canberra.

Prasad, S., Langridge, P. 2012, 'Australia's role in global food security', *Office of the Chief Scientist Occasional Paper Series* 5.

FOOD FOR THOUGHT

GM is fast and hi-tech but humans have been
changing crops for millennia, writes **Peter Langridge**

N 2002, THE Jewish Orthodox Union concluded that genetically modified (GM) foods, including those with pig genes, may be consumed by Jews. The Islamic Jurisprudence Council followed suit. After similar debate, it decided foods produced from genetically manipulated crops are halal and suitable for Muslims, although there may be issues with crops that contain DNA from forbidden foods.

Today's concerns about the influence of genes on the nature and characteristics of food crops would no doubt surprise ancient farmers. Not so much by the act of manipulating plants but by the speed of the technology. After all, since the birth of agriculture around 10 000 years ago farmers, and more recently breeders and scientists, have sought to increase the efficiency, reliability and safety of food production.

Early farmers selected varieties, or lines, that allowed them to improve production systems and support harvesting and storage. This led to major changes in characteristics of plants compared to their wild relatives. Importantly, farmers selected plants that performed well when grown as a community, a crop. Farmers in different regions continued to select plants for their environment and over time diverse landraces were generated.

Farmers also exchanged lines with neighbours and along trading routes, creating a flow of genetic material over regions and continents. In each area, farmers continued selection for the best-performing lines. For instance, when Europeans arrived in Australia they brought their crops with them. Many performed badly, so selection for lines adapted to our environment became a priority.

FROM FIELDS TO LABORATORIES
Selection by farmers underpinned modern agriculture but much higher rates of genetic improvement could be achieved by systematic breeding. The discovery of the principles of genetics laid the foundations for rapid improvements in crops over the past century. As knowledge of genetics and genes has expanded the rate of crop improvement has accelerated.

Selecting the best
Systematic breeding uses variation to create new gene combinations. For example, a wheat variety may show high yields but be susceptible to a harmful disease. The breeder can correct this defect by crossing to a plant that has resistance to the disease but may be low-yielding. Offspring of the cross have different combinations of genes from parents and are screened to identify the individuals that have inherited both the good yield and the disease resistance. Breeders select the best-performing plants to continually improve yield and quality of their crops. The more diverse the variation available, and the more plants the breeders can screen, the greater the opportunity for advances.

Selective breeding is essentially a numbers game since many important crop characteristics, such as yield, drought tolerance and fruit size, are complex and controlled by a very large number of genes. This means that finding the best combination of genes can be difficult. However,

mechanised sowing and harvesting now enables breeders to grow and assess thousands of genetic combinations. Additionally, the advent of computing and powerful statistical methods supports the accurate evaluation of performance of the new plant lines.

More recently, DNA markers have allowed breeders to follow individual genes as they are passed to the offspring. DNA markers are like DNA-fingerprinting used in forensic science and allow breeders to track desirable genes, such as for disease resistance or larger fruit, in

From a scientific perspective, genetic engineering is a logical development and refinement of earlier techniques.

the many thousands of plants in their programs. Based on the DNA fingerprint, the breeders can predict many of the key characteristics of the plants when grown as a crop, such as disease resistance, quality and even yield.

Generating new genetic variation
Breeders continually search for new methods for expanding the number of useful genes available for cross-breeding and selecting in the next generations of plants. The search for new variation has taken breeders well beyond the traditional view of breeding as crosses between sexually compatible plants. In the 1950s chemicals or ionising radiation were widely used to produce mutations and induce novel changes. Many modern varieties of crops and fruits, such as several seedless mandarin varieties and many new disease resistances include these mutations.

Return to the wild side
Breeders have also made extensive use of the wild relatives of our domesticated plants by methods known as wide-crossing. Genetic information from the wild relatives is integrated into the genetic make-up of the crop plant by a process called chromosome engineering.

Currently, the most widely used genes in genetically engineered crops provide tolerance to herbicides or resistance to insect pests or viral diseases

This form of engineering has provided a key technique for improving a range of characteristics, particularly disease resistance.

These methods involve large-scale changes in the genetic make-up of crop plants and bring hundreds or thousands of new genes into the crop. In both chromosome engineering and mutation breeding, thousands of genes are altered and most of the changes will actually reduce the performance of the plants. Scientists and breeders clean up the genetic material by slowly removing unwanted, often deleterious genes or mutations, while retaining the desirable genes.

These techniques have been used for half a century and have been critical to the success of current crops such as wheat, corn, rice, barley, and many fruits and vegetables. Importantly, the new plants produced by chromosome engineering or mutation breeding are not subject to regulation and can be generated, screened and commercialised without the need for special scrutiny by government and international regulatory agencies.

Genetic engineering

From a scientific perspective, genetic engineering is a logical development and refinement of earlier techniques. Most plants and animals contain around 30 000 genes. A major advantage of genetic engineering is that single genes are modified, not hundreds or thousands, and modifications are specific for the desired feature, such as disease or pest resistance.

A genetically engineered plant contains one or more genes that have been inserted into the genetic make-up of the plant using recombinant DNA technology. This technology involves isolating or synthesising a gene in a laboratory and then transferring it to the new host plant. Once inserted into the genetic make-up of the host plant, the gene will be passed on to its descendants in the same way as all the other genes that make up the organism. The genes can be from any source but they may need to be modified to work properly in their new host. The most widely used genes in genetically engineered crops currently being grown provide tolerance to herbicides or resistance to insect pests or viral diseases.

Genetic engineering has developed into a major scientific tool also supporting many aspects of biological and medical research. Virtually every university and biological research laboratory routinely engineers bacteria, plants or animals in the course of research. The result: over the past few decades knowledge and understanding of genes, their function and regulation has exploded.

THE UPSIDE

As understanding has grown, scientists are increasingly confident in the viability and safety of genetic engineering. However, regulatory complexity limits development and delivery of practical outcomes for this technology.

For years, genetic engineering has been seen as the future direction for crop improvement. Where available it's rapidly adopted by growers. Over the past few years the area sown with genetic-engineered crops – GM crops – in developing countries has outstripped production in advanced economies. The traits so far have improved production and provided benefits for farmers. However, GM plants with improved nutritional characteristics such as high vitamin A, high iron and resistant starch have been developed and are close to release.

Many public and private sector groups are evaluating GM crops for improved drought and salinity tolerance. Importantly, GM technologies could offer greater flexibility in addressing challenges resulting from increasingly variable climatic conditions and the need to reduce the use of pesticides, fungicides and fertilisers.

The United Nations Food and Agriculture Organisation estimated in 2012 that wheat yields must be increased by 70 per cent by 2050 to meet expected demand. All technologies, including GM, will be needed to meet the target.

THE DOWNSIDE

Many concerns about GM technology relate to ethical and political issues. Some of the first GM plants contained not only the genes

GM RICE TO TACKLE NUTRITIONAL DEFICIENCIES AFFECTING MORE THAN TWO BILLION PEOPLE

According to the World Health Organisation, iron deficiency is the most common and widespread nutritional disorder in the world and affects more than two billion people (30 per cent of the world's population). Symptoms range from poor mental development and depressed immune function to anaemia.

A team of scientists in the Australian Centre for Plant Functional Genomics, led by Dr Alex Johnson, have engineered rice to contain four times more iron than found in conventional rice. This was achieved by boosting the expression of a naturally occurring rice gene so that the plants respond as if they were short of iron. Rice is the main source of food for almost half the world's population, particularly in developing countries, and the levels in the GM lines is high enough to meet the daily recommended requirements.

The GM rice lines have been tested in field trials in South America and will now go into trials in Asia.

for the new trait but also genes for resistance to antibiotics, as can happen when genes are transferred from a bacterium that possesses antibiotic resistance. This helped scientists select for plants that had been engineered but raised concerns that the antibiotic genes could exacerbate the problems of antibiotic-resistant bacteria. However, several studies showed that the genes were not transferred from the plants to bacteria and scientists soon developed methods to remove the antibiotic genes from the GM plants.

Opponents of GM technologies also raised questions about the safety of foods produced from GM crops. The European Union spent more than half a billion dollars on studies of GM safety to support the political anti-GM stand, only to conclude that GM-derived foods are as safe as conventional crops. Unfortunately, these safety studies are dismissed by anti-GM groups.

Still, there are concerns about corporate control of the technology. The GM crops currently grown were all released by large multinational seed companies. Ironically, high regulatory costs strengthen the corporatisation of food production and weaken public sector engagement by making it virtually impossible for public-sector organisations to take a GM technology through to farmers' fields.

Scientists are increasingly confident in the viability and safety of genetic engineering.

A POLITICAL BOTTLENECK?

Almost three decades of plant biological research was based on the assumption that GM technology would be routinely applied. But delivery of these outcomes is slow and costly. Consequently, major investment is flowing into finding alternative techniques that won't be classified as GM. In my opinion this risks further polarising the European/American divide as North America will likely classify new techniques as non-GM while the Europeans will continue GM-style regulation.

GM technology is espoused as a critical tool for ensuring global food security. However, it's also a source of dispute in agricultural aid programs with several European countries tying aid to non-GM approaches. Pressure is also on developing counties to take on the regulatory regimes of developed countries that are either strongly pro- or anti-GM, rather than considering local needs and capabilities.

Humanity enjoys the safest and cheapest food in history – and prehistory – through the industrialisation of food production. Simultaneously, our population is increasingly urbanised and poorly informed about modern agriculture. Little wonder many people reject industrialised production systems in favour of more traditional methods, while still demanding cheap, safe food.

I argue this cannot be done without using new scientifically validated technologies. Scientists must work to gain community support through open discussion. We can be sensitive to the concerns of consumers and conscious that not all concerns are related to the science; we can help the Rabbinical and Islamic Jurisprudence Councils by avoiding pig genes in our crops. It's a tough task, an ongoing task, but for me it's a critically important task.

PROFESSOR PETER LANGRIDGE is a University of Adelaide plant scientist and the CEO of the Australian Centre for Plant Functional Genomics.

FURTHER READING

Acworth, W., Yainshet, A. and Curtotti R. 2008, 'Economic impacts of GM crops in Australia', research report, Department of Agriculture, Forestry and Fisheries, Australian Government.

Cormick, C. 2007, 'Public attitudes towards GM crops and foods', *Agricultural Science* 21(2): 24-30.

European Commission 2010, 'A decade of EU-funded GMO research (2001-2010)', http://ec.europa.eu/research/biosociety/pdf/a_decade_of_eu-funded_gmo_research.pdf.

Snell, C. et al 2012, 'Assessment of the health impact of GM plant diets in long-term and multigenerational animal feeding trials: a literature review', *Food and Chemical Toxicology* 50: 1134-48.

WHITHER AUSTRALIA'S WATER

Climate change is disrupting the ebb and flow of the continent's water supplies, writes **Åsa Wahlquist**

Although Australia is the driest inhabited continent, under the ground and out of sight it possesses vast reserves of water. Groundwater supplies about 17 per cent of the water the nation uses. Many areas depend on it.

ON MELBOURNE CUP Day, 2006, Prime Minister John Howard called a meeting to discuss the unprecedented drought in the Murray-Darling Basin. Afterwards, he commissioned the biggest single project the CSIRO has ever undertaken: the Murray-Darling Basin Sustainable Yields Project.

It has since been extended to other areas including northern Australia and the Great Artesian Basin. The resulting world-leading expertise is about to be exported, to assist nations from the USA to India to understand and manage their water better.

Dr Bill Young, director of CSIRO's Water for a Healthy Country Flagship, says Australia now has "a much more robust picture of how much water we have, its variability in space and time, and the likely impacts of a changing climate on both of those things".

Australia truly is a land of droughts and flooding rains. Not only does it have an extraordinarily variable rainfall – the long dry in the Murray-Darling Basin ended in 2010 with two record-setting years for high rainfall – but climate change is now affecting the rain as well as inflows into rivers and dams.

But while the east coast flooded, the south-west became even drier. That dryness is directly attributed to worsening climate change, and Perth's water managers are planning for a future with no dam water.

One of the great ironies is that although Australia is the driest inhabited continent, under the ground and out of sight it possesses vast reserves of water. Groundwater supplies about 17 per cent of the water the nation uses. Many areas, particularly in inland Australia, depend on it.

Groundwater is a finite resource, replenished only when surface water seeps into the aquifer. There are fears that some aquifers have reached 'peak water', as it is being used faster than it's recharging.

Recharge can be a slow process. Some of the 64 000 million litres of water held in the Great Artesian Basin (GAB), which lies under one-fifth of Australia, is 2 million years old (see: **Figure 1**).

Young, a natural resources engineer, admits the GAB "is a lot more complex

Wivenhoe Dam, important to Brisbane's water supply, is subject to droughts and flooding.

than it was thought". The CSIRO and Geoscience Australia have produced stunning 3D images to help communicate the new understandings. One thing they didn't expect is that the GAB has different levels of connection with nearby basins.

CSIRO research scientist, hydrogeologist Dr Brian Smerdon, says while the eastern side is predicted to have increased groundwater levels, the western side is likely to have lower levels, due to a very long-term natural decline.

"It is very difficult to know the precise rates of inflow and outflow for such a complex groundwater basin," Smerdon says. "It appears that outflows are greater than inflows for most of the Basin. Continued extraction of groundwater from the Basin requires continued measurement of the groundwater levels."

Groundwater is frequently linked to surface water. As hydrologist Richard Evans points out, this makes it possible to "allocate the same resource twice, to surface-water users and to groundwater users". In a number of catchments in the Murray-Darling, groundwater extractions exceeded recharge.

THE SOUTH-WEST'S HOTTER, DRIER FUTURE

In the mid-1970s, the south-west of Western Australia began to dry. Rainfall declined and inflows into dams fell even more dramatically. Between 1911 and 1974, the inflow into Perth's dams averaged 338 gigalitres (GL) or billion litres. By 2006 to 2010 it was down to 57.7 GL. Today Perth relies on dams for 31 per cent of its water and on desalination for 23 per cent while groundwater supplies the remainder.

Dr Don McFarlane is the WA coordinator for the CSIRO's Water for a Healthy Country flagship. He says rain that does fall no longer delivers the run-off into dams it once did. "It is not just the drying of the catchment. The groundwater systems have now fallen in some cases over 10 metres." That disconnected groundwater and surface water. "We think because the global climate models are all projecting a hotter and drier future that this is an irreversible change," McFarlane says. "The Water Corporation expects they will have some years when they won't have any useful run-off into the dams."

CSIRO scientist Dr Wenju Cai says this decline in rainfall is one of the most compelling examples of human-induced climate change in the world. He explains the system that brings rain to WA during winter has moved towards the South Pole. The poles are warming faster than other latitudes, reducing the temperature gradient between the poles and the equator. This has led to a major change in global atmospheric circulation and a decrease in the subtropical jet stream and a decline in winter storm tracks over the south-west.

According to McFarlane, Perth lies over several major aquifers. The city is currently pumping from a very deep aquifer which is replenished slowly. "The amount of water they are taking out is far in excess of the recharge," he says. Pumping has at least reduced water pressure, "by making room in the aquifer they are sucking water in. It is still not enough," McFarlane says.

Perth does have a water future, but it is one that will be increasingly reliant on cutting-edge science, on desalination and recycling and a better understanding of its underground water.

DREAMING OF NORTHERN WATER

Australians have frequently looked north and fantasised about using the heavy monsoonal rains that fall there.

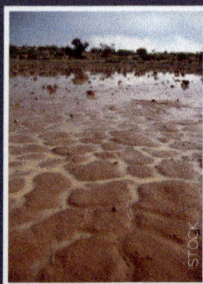

But the CSIRO Northern Australia Sustainable Yields Project found the region is in fact seasonally water limited because of the potential water loss from vegetation and from evaporation far exceeds rainfall. Rainfall is highly seasonal: 94 per cent falls in the wet season and the remaining three to six months of the year are dry and the climate is extreme and changing.

Most of the rain falls on the flat coast, where there are limited opportunities for dam-building. The best dam-building sites are in the upper reaches of the catchments where rainfall is sporadic and evaporation highest.

The Northern Australia Taskforce found groundwater development provided the best prospect. It noted the only large northern irrigation development, the Ord River Scheme, opened in 1972. Unfortunately, the wet and humid conditions have proved perfect for pests and diseases, while distance and lack of infrastructure has hindered exports. More than $500 million has been invested in the scheme by the WA and federal governments so far, yet it currently produces just $140 million in agricultural products, including rockmelons, pumpkins, mangoes, chickpeas and chia. The main crop is sandalwood.

According to Dr Bill Young, the CSIRO is undertaking a major project in the north "where we are testing soils, looking at the terrain and working out where you could put dams that would offer a good water yield".

He argues Australia should develop its northern water resources, but it must be done "in a careful, reasoned and evidence-based way, and not repeat the mistakes of the Murray-Darling. We have the wherewithal to do the science and do it in a smart way and then it will be a sustainable development in the nation's interest."

Figure 1: A visualisation of the Great Artesian Basin (GAB) showing the depth to the water table (depth in colour scale, with a positive down direction). Watercourses are shown as blue lines. Extent of the GAB is the red line. State, territory and coastal boundaries are shown as black lines. Vertical exaggeration is x50.

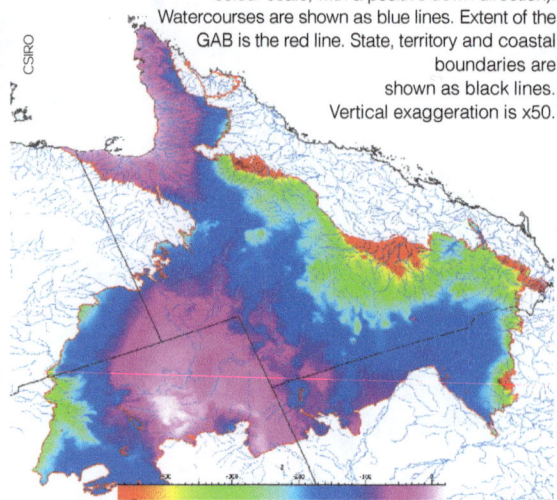

In 1972, drought hit eastern Australia. Monash University climatologist Professor Neville Nicholls, then with the Bureau of Meteorology, says: "We really didn't know it was an El Niño until 1973, after it had gone. By 1982 we had some real-time monitoring capability and by 1997 it was terrific." Identifying the El Niño-Southern Oscillation (ENSO) has been a significant step in understanding, and managing, Australia's waters.

Just how much rain falls on Australia depends on how much evaporates from the sea surface and where the weather systems deliver that moisture-filled air. In El Niño events, the warm water and clouds move away from Australia, often, but not always, bringing drought to eastern Australia. In La Niña years, the warm water and clouds move to the western Pacific, and trade winds blow the clouds over eastern Australia where they yield rain, as they did so spectacularly in 2010 and 2011.

In 1999, climate scientists identified a similar pattern in the Indian Ocean, called the Indian Ocean Dipole (IOD). CSIRO Marine and Atmospheric Research scientist Dr Wenju Cai says the frequency and intensity of the north-west cloud band events that bring rain are suppressed during a positive IOD. They are often, but not always, correlated with an El Niño.

The Norman River, which runs into the Gulf of Carpentaria, is a flood zone in Far North Queensland.

"South-eastern Australia, the Murray-Darling Basin, is more controlled by the positive IOD," Cai explains. "The Indian Ocean Dipole we now come to understand is perhaps even more important than ENSO." (See: **See-saws and boys and girls**, in Marine Life In A Changing Climate, p21, for more detail.)

This improved understanding should lead to more accurate seasonal forecasts. "The science is going to fill the gap," Cai says. "We know the ocean variability, sea-surface temperature is very important. Predicting sea-surface temperature, we get a good handle on that, but from sea-surface temperature to rainfall is the hardest bit, because it requires a climate model to simulate convection properly."

Young would like to see more work done on city water supplies, looking at climate change, supply and demand "in the context of a growing and urbanising population, and work out what that means for the right investments in water infrastructure in terms of centralised and decentralised systems". He points out our cities are growing and urbanising, and wants good science to inform both large- and small-scale investments in urban water infrastructure.

Urban Australians proved their water awareness during the dry of the 2000s, adopting water-saving technologies, obeying water restrictions, and in Melbourne and Brisbane in particular, seriously reducing their household water use.

Young says the science that has developed out of the crisis in the Murray-Darling has resulted in Australia becoming a world leader. With the world's demand for food rising, access to water will be critical. Young says Australian scientists will be at the forefront "exporting our expertise both in a science and technology sense and in a business sense to help address some of the global emerging problems".

ÅSA WAHLQUIST is a rural writer.

FURTHER READING

CSIRO Sustainable Yields Projects – a comprehensive scientific assessment of current and future water availability in major water systems across Australia to provide a consistent framework for future water policy decisions.

Groundwater Essentials, National Water Commission – The National Water Commission is responsible for driving progress towards the sustainable management and use of Australia's water resources under our blueprint for water reform – the National Water Initiative.

Great Artesian Basin – This CSIRO-led project examined the water resources of the Great Artesian Basin and assessed potential impacts of climate change and resource development.

Perth rainfall, dam storage and water supply measurements, http://www.watercorporation.com.au/water-supply-and-services/rainfall-and-dams.

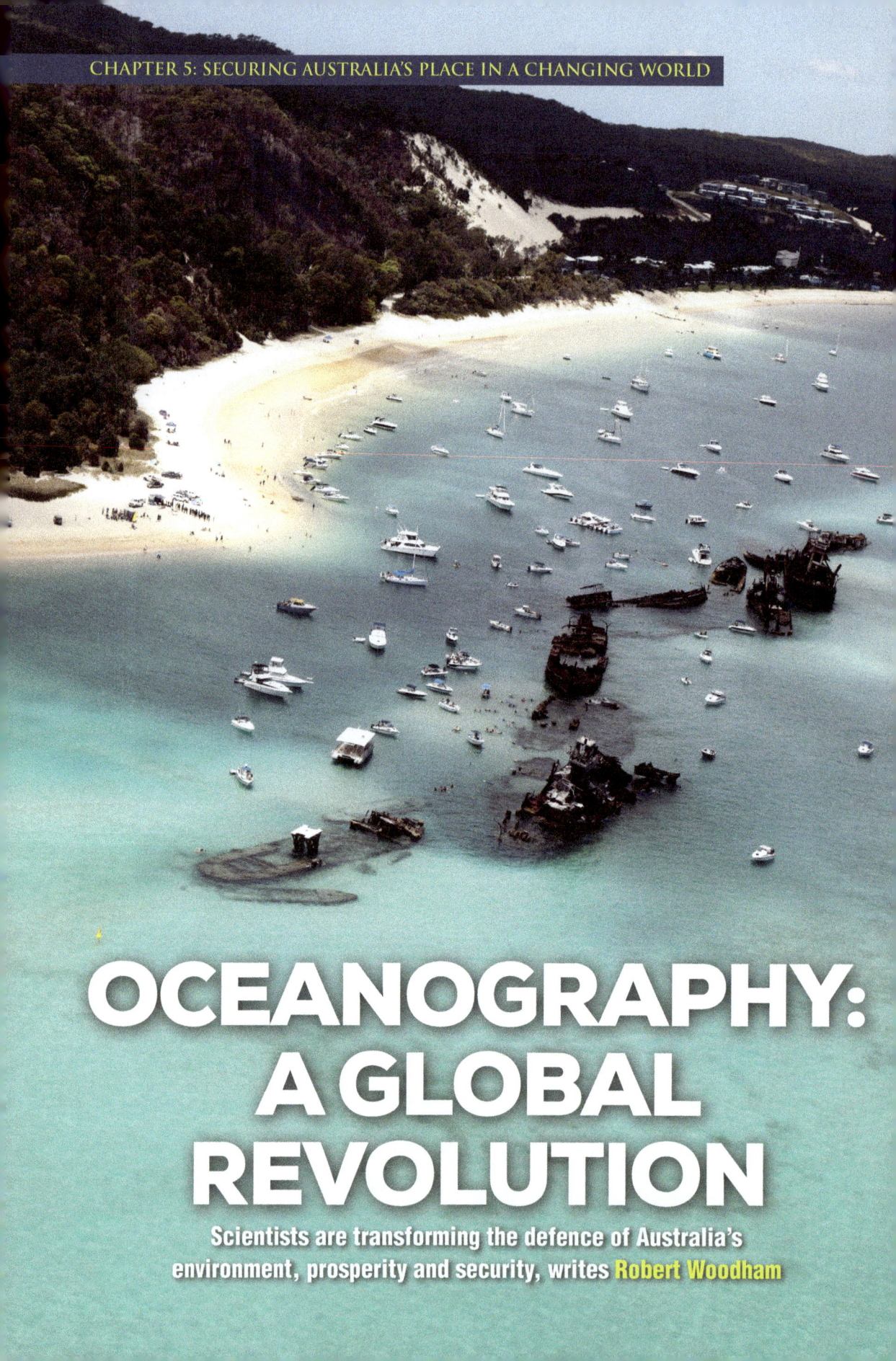

OCEANOGRAPHY: A GLOBAL REVOLUTION

Scientists are transforming the defence of Australia's environment, prosperity and security, writes Robert Woodham

ON 11 MARCH 2009, the 24 000-tonne container ship *Pacific Adventurer* was damaged in heavy seas whipped up by tropical cyclone 'Hamish' which had been moving southwards through the Coral Sea over the preceding week.

As the ship floundered off the Queensland coast near Point Lookout, on the north-east corner of North Stradbroke Island, about 270 tonnes of oil spilled into the ocean and 31 shipping containers full of ammonium nitrate were lost overboard.

Not only did the oil pose a major threat to marine life, the ammonium nitrate was explosive. It might also fertilise an environmentally damaging algal bloom. Moreover, the containers themselves posed a shipping hazard.

The Queensland Government, assisted by federal agencies, reacted quickly to the disaster but was faced with two difficult questions: where would the oil slick end up? And where exactly had the containers gone?

Both these questions were answered with the help of an innovative new ocean forecasting system called BLUElink, developed by a partnership between the Australian Bureau of Meteorology, CSIRO and the Royal Australian Navy.

The Navy needed the system to provide forecasts of oceanographic conditions that might affect maritime warfare such as hunting for submarines, mine warfare and amphibious operations. But as it despatched its state-of-the-art minehunters, HMA ships *Yarra* and *Norman*, to Point Lookout to help locate the missing containers using their advanced minehunting sonars, Navy oceanographers were busy interpreting the BLUElink forecasts for a more peaceful application.

The ships' crews needed forecasts of the surface and subsurface currents in order to deploy their variable depth sonar and remotely operated vehicles safely and also to understand how the acoustic properties might affect sonar performance.

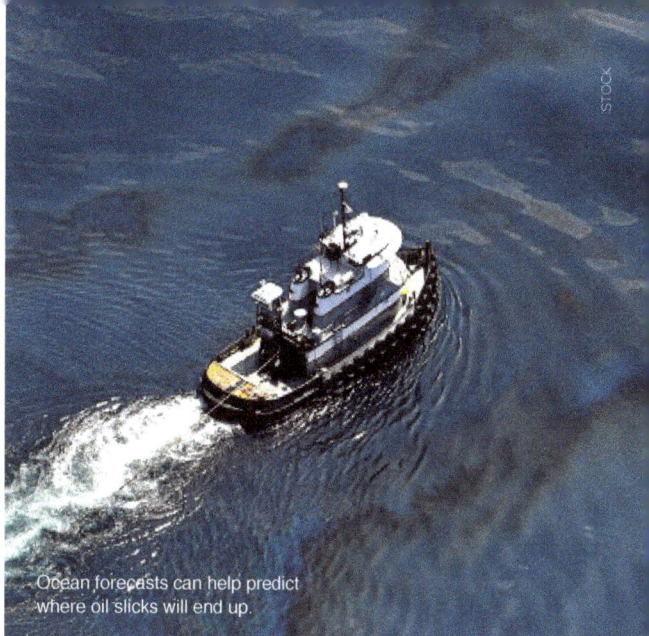

Ocean forecasts can help predict where oil slicks will end up.

Meanwhile, Gold Coast oceanographer Brian King was tasked to advise on the likely trajectory of the oil and ammonium nitrate spills. Which beaches were under threat?

Using BLUElink ocean forecasts, as well as comparison forecasts from other systems, he identified a stretch of coast between Coolum Beach in the north to Bribie Island in the south, as well as the north-eastern coastline of nearby Moreton Island itself, as the areas most at risk.

Over the following days, the Navy minehunters located all 31 of the missing containers. The spill response, involving around 2500 people, started without delay as the oil came ashore exactly as predicted. BLUElink had proved its worth.

So what exactly is an 'ocean forecast' and why is there a rapidly growing need for them in the 21st century?

Most people are familiar with the depressions, anticyclones, fronts and trough lines that bring Australia its varied weather, our land "of droughts and flooding rains". Perhaps not quite so obvious, though, is the fact that Australia's climate variability is matched by the enormous complexity in the surrounding seas.

The East Australian Current (EAC) flows down the east coast then deflects offshore, shedding warm- and cold-cored eddies into the Tasman Sea, while the Leeuwin Current flows down the west coast on an annual cycle starting in late summer. The Antarctic Circumpolar

Current flows strongly and constantly eastwards to the south of the Australian continent and the Indonesian Throughflow affects waters to the north (See: **Figure 1**).

In addition, numerous other oceanographic phenomena such as upwelling of cold water to the surface, internal waves and extreme tides are found in various locations around our waters.

These features are defined by their temperature, salinity and current structures, all of which are forecast by the BLUElink system in the same way that the weather is forecast by numerical models that represent the temperature, moisture and winds in the atmosphere.

The motivation for ocean forecasting has grown in recent years, as experts realise the importance of the ocean to the weather systems such as El Niño, the extensive warming of the central and eastern tropical Pacific that leads to a major shift in weather patterns across the Pacific. El Niño events often mean cooler than normal sea surface temperatures in the western Pacific and an increased probability of drier conditions in Australia.

The 1982-83 El Niño was the strongest on record at the time, but despite the havoc it caused around the Pacific Rim, the weather phenomenon was not even detected in the ocean until it was nearly at its peak. This prompted a concerted international effort to improve the poor state of ocean observations.

As a direct result of the 1982-83 El Niño, a fixed array of buoys was installed in the equatorial Pacific Ocean to monitor the three-dimensional temperature structure and, hence, provide warning of future events. Satellite observations of sea surface height that show the locations of oceanic eddies have been improved, along with measurements of sea surface temperature and other oceanographic properties.

More recently, there have been various initiatives to improve measurements in the ocean.

One such initiative is the 'Argo' program, which has put in place a global fleet of over 3500 floats to observe ocean temperature, salinity and currents.

Argo floats loiter in the deep ocean, but every 10 days they are programmed to rise to the surface, measuring temperature and salinity as they go. At the surface the buoys transmit their data over a satellite link to shore-based data centres where it is shared around the world. Australia has been a particularly strong supporter of Argo, contributing around 50 floats a year, some of which have been deployed from Navy ships.

Now the ocean is being observed much more comprehensively, the large volume of data collected each day can be used to create a snapshot of ocean conditions. This snapshot is used as a starting point for the ocean-forecasting models now being set up in centres around the world. Although this revolution in so-called operational oceanography is just getting going, Australia is at the forefront, notably through the BLUElink project.

The RAN has not been slow to see the opportunities offered by operational oceanography, particularly in view of the complex waters in our region. These are waters in which the navy may one day have to fight – hence its involvement in BLUElink. As a result, the Navy is in the very top rank of world navies in terms of oceanographic capability.

The Navy is interested in various types of oceanographic forecasts. Sound speed in the ocean varies according to temperature, salinity and pressure, and these variations cause sound to bend or 'refract'. This can help submarines to camouflage the sound that might give them away.

For example, a submarine might hide in the shadow zones caused by refraction beneath the

> The 1982-83 El Niño was the strongest on record at the time, but despite the havoc it caused around the Pacific Rim, the weather phenomenon was not even detected in the ocean until it was nearly at its peak.

BLUElink data helps locate shipwrecks, including AHS *Centaur*, which was torpedoed off the coast of Queensland in 1943.

mixed surface layer or close to an eddy boundary. In contrast, a surface ship or maritime patrol aircraft might try to exploit a sound channel, where refraction concentrates sound energy, in order to detect a submarine at long range.

Oceanographic forecasts are also useful inshore. That's because mine warfare specialists and amphibious forces need to know about the currents, acoustic properties, swell and surf conditions, longshore currents and rips. Ocean currents, at and below the surface, can also be exploited by surface ships and submarines to increase their speed over the ground.

The case of the *Pacific Adventurer* illustrates a peaceful application of oceanographic forecasts, but there are many others. For instance, BLUElink data guided the search for HMAS *Sydney II* and the German cruiser *Kormoran*, which sank after a sea battle in 1941, and also for AHS *Centaur*, torpedoed off the coast of Queensland in 1943. All these wreck sites were located by international shipwreck hunter David Mearns and his team, helped by BLUElink.

In addition to improving management of the

seagoing disasters and assisting underwater explorers, BLUElink data can improve the handling of oil spills. The system, for example, was used to plan the response to a major oil spill in the Timor Sea from the 'Montara' oil platform after a well-head blowout in August 2009.

The applications of BLUElink and systems like it keep growing. The data they provide is invaluable to university and research institution scientists, seeking to understand a wide range of ocean phenomena. Ocean engineers in the oil and gas industry also exploit such data to plan ocean renewable-energy projects.

National weather agencies are interested in using ocean forecasts to increase the skill and range of weather forecasts by modelling interactions between the ocean and the atmosphere. The *Marine Nation 2025* report, published in March 2013 by the Oceans Policy Science Advisory Group, points out that the ocean contributes approximately $44 billion per annum to our economy though a range of marine industries. This is expected to increase to $100 billion by 2025, as these industries expand,

The structure of some of the ocean currents around Australia, as revealed by BLUElink forecasts of sea surface temperature and sea surface height, for 1 June 2013.

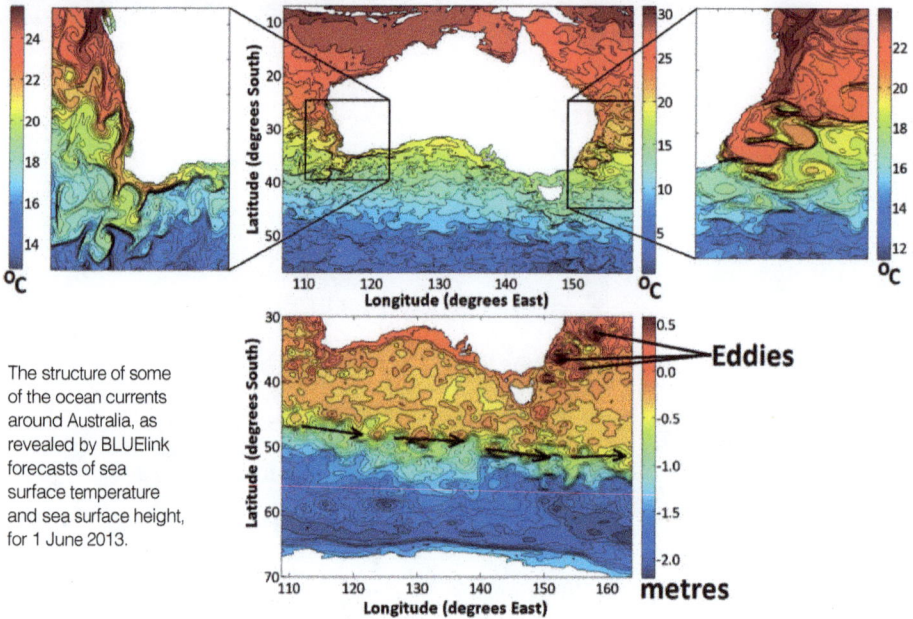

Figure 1: BLUElink ocean current forecast for 1 June 2013

and new initiatives in offshore oil and gas, ocean renewable energy and fisheries come on line. This growth will drive an increasing demand for oceanographic data.

Clearly, a global revolution in operational oceanography is in full swing. This international scientific transformation is driven by an increased awareness of the effects of the ocean on our climate, as well as increasing economic activities at sea and the international sensitivities posed by competing interests.

Australia is very much at the forefront and little wonder. The oceanography around our island nation is complex and provides scientists with an enticing ocean laboratory. The BLUElink

ocean forecasts offer a decisive edge to the Navy's fighting ability, and are starting to reveal some of the mysteries and complexities of our ocean environment. They will also help mitigate future environmental threats, such as the *Pacific Adventurer* disaster.

COMMANDER ROBERT WOODHAM is a Meteorology and Oceanography (METOC) specialist in the Navy. He holds a Master of Science degree in applied oceanography, and is studying for a PhD in oceanography. Currently Assistant Navy Scientific Adviser, he is a former Navy Director of Oceanography and Meteorology.

FURTHER READING

ATSB Transport Safety Report, Marine Occurrence Investigation number 263, 'Independent investigation into the loss of containers from the Hong Kong registered container ship Pacific Adventurer off Cape Moreton, Queensland 11 March 2009', http://www.atsb.gov.au/media/2905740/mo2009002.pdf.

Royal Australian Navy, 'Minehunter, coastal (MHC)', www.navy.gov.au/fleet/ships-boats-craft/mhc.

Catalyst 2010, television broadcast, ABC TV, 24 June 2010, http://www.abc.net.au/catalyst/stories/2936244.htm.

Bureau of Meteorology, 'Sea temperature and currents', http://www.bom.gov.au/oceanography/forecasts/index.shtml.

Latest data from the Tropical Atmosphere Ocean (TAO) array, which monitors the Pacific Ocean for the prediction of the El Niño/Southern Oscillation, http://www.pmel.noaa.gov/tao/

Information on Argo profiling floats, http://www.argo.ucsd.edu

Mearns, D.L. 2009, *The Search for the Sydney: How Australia's greatest maritime mystery was solved*, HarperCollins, Sydney.

The Oceans Policy Science Advisory Group 2013, 'Marine Nation 2025: Marine science to support Australia's blue economy', OPSAG report March 2013.

COMPUTER CRIME IS ON THE RISE

Cybersecurity is an issue for everybody, from individuals to nations, writes **Alastair MacGibbon**

I RECEIVED THE SAME email five times in as many weeks informing me of an $18.60 refund following a 'billing error' with a 'mobile phone provider'. Not a huge sum, but believable.

I don't have a mobile phone with this company, so I ignored it the first time, and the second.

But by the third email I had started to wonder: did someone else in my family have a phone with this company? And by the fifth email it took considerable strength not to click on the attachment. After 15 years policing experience and more than 10 years addressing cybercrime, I was tempted. Scary. What if it was a legitimate email communication?

I checked the full email header, rather than the summary from/to/date header we see by default, and noticed the email originated in the United States, not Australia. This and a visit to a scam alert website that noted this particular email restored faith in my initial gut judgement, allowing me to rest easy knowing that I wasn't missing out on the big bucks.

The emails were nothing more than spam. Determined, simple spam. Had I filled out the attachment I'd have entered personal information, including financial, so the 'refund' could be paid.

Had my credit card details been successfully obtained to use fraudulently, I would be reimbursed by my bank under the terms of the e-Payments Code. So, no real loss. Or is there? Unsurprisingly, the banks don't take one for the team in the online economy.

Their moment as the 'white knight' sees them demand reimbursement from the online merchants where stolen credit card details were misused. 'Card not present' – internet and phone – transactions are at the merchant's risk. In turn, merchants build that loss into the cost of goods and services legitimate customers buy. So everyone loses out, bar the criminal.

Back to my original series of spam scam emails: the criminal also could have embedded a 'malicious payload' of computer code in the attachment known as 'malware'. A 'key logger'

may have been installed, sending every stroke I type – including passwords – back to the criminal. My contacts and files may have been plundered and abused by the criminal then resold to other criminals via well-developed criminal online black markets.

My computer – now made a 'zombie' by the malware – could be drafted with other compromised computers into a 'botnet army' and used to carry out denial-of-service attacks against business and government websites, causing them to crash. My computer may also have been used for breaking, or hacking, into other computers with all roads leading back to me if the matter was investigated. My computer could store and share illegal material such as child

The sheer scope of cyber vulnerabilities alone helps make a compelling case for national security concern.

WORRISOME WORMS

In 1999 the 'Melissa' worm, and then in 2000 the 'ILOVEYOU' worm, provided early warnings of the speed with which malicious codes can spread across connected computers. These worms – a form of malicious computer code that spreads across networks – were particularly significant because they impacted business, government and consumer systems. Fortunately, while both were annoying and caused some harm to targeted computers they were reasonably harmless.

Ten years later, when security researchers learned of the 'Stuxnet' worm, the world of large-scale computer malware had changed. Stuxnet – widely believed to have been developed by the US intelligence community in cooperation with Israel – was originally designed to target Siemens supervisory control and data-acquisition systems used in Iran's uranium-enrichment program. The code caused centrifuges to spin out of control, damaging the country's nuclear efforts. Although specifically targeted, Stuxnet spread and was later altered and used for other malicious activity.

Subsequent incidents have occurred, either carried out by or in support of nation states, utilising malicious code. As well, successful criminal malware enterprises such as 'Zeus' have targeted banking information.

pornography and could be used to send more spam just like the one that caused all my trouble.

Maybe all of the above, over time. A tempting $18.60 would have reaped significant individual loss, while contributing to a burgeoning criminal economy and supporting infrastructure. This scenario is played out in an automated, relentless fashion every second we're awake and asleep.

And the increasing use of mobile and tablet devices, combined with a steady growth in online activities, multiplies the threats. For instance, what if the compromised computer, phone or tablet device I use for my personal life is also used for work? The corporate system could also be at risk, exposing company intellectual property, client information, finances and more.

Ultimately, what was first a damaging individual incident – when aggregated with potentially thousands or indeed millions of other individual incidents – could have national security implications, threatening Australia's economic interests, the wellbeing of the Australian public and the integrity of Australian Government information and systems.

The sheer scope of cyber vulnerabilities alone helps make a compelling case for national security concern. In a recent example, a seemingly benign hacker nicknamed Carna compromised 420 000 internet-connected devices, mainly routers and servers, to create his own botnet. While Carna claimed to have

Cyber vulnerabilities mean computer systems holding valuable information or running critical infrastructure can be targeted and damaged by criminals.

no malicious intentions the incident illustrates the potential size of internet security issues.

Malicious botnets like Waladec and Rustock – successfully crippled by Microsoft's Digital Crimes Unit in 2011 – and the more recent Bamital botnet are examples of highly malicious criminal enterprises that affected hundreds of thousands of people worldwide.

Cyber vulnerabilities at a small business, corporate and government level mean that valuable intellectual property and traditional national security secrets can be targeted, as can computer systems running critical infrastructure supporting the economy: power, water, transport, food distribution, telecommunications and banking.

And in some instances that targeting may have found its mark via, say, a scam mobile phone refund email.

To date, we've failed to grasp the enormity of the misuse of technology and, as a result, have not viewed the problem as a societal one. The $18.60 refund scenario highlights how cybercrime is both an individual and national security issue. Cybercrime can be so interlinked that, theoretically, my $18.60 click could be part of a larger, orchestrated attack on critical infrastructure. It's essential

to recognise that no matter how benign a scam may seem it is potentially malignant and can definitely metastasise.

When addressing these issues, blame is usually attributed to end users or government agencies, particularly security services and police. There are few calls for internet service providers, online retailers, social network operators, software and hardware manufacturers and businesses in general to shoulder greater responsibility in providing safer services and educating end users.

True, end users and governments must scale up their efforts. But what's needed most is a national approach addressing cybersecurity like a public health concern: with measurable baseline data, broad strategies and a relentless long-term commitment to tackle the problem.

Scientists, engineers and mathematicians can and should play a central role. Instead, a handful of public officials and information technology (IT) security professionals dominate the debate.

In the age of the internet, the once-dominant 'three Rs' of reading, writing and arithmetic have been replaced by the 'three Cs': coding, computation and communication. Consequently, Australia requires more

DIGITAL IMMUNE SYSTEMS

Over the past 20 years, teams of US cyber specialists have worked on an innovative approach to cybersecurity originally proposed in 1992 by scientists at the University of New Mexico and the Los Alamos National Laboratory. Drawing inspiration from human immune systems, the goal is to create digital immune systems and healthy computer ecosystems able to defend against cyber threats.

Computer viruses are similar to biological viruses. They disrupt healthy systems and exploit their hosts in order to replicate. While mimicking human immunity makes sense, the human immune system is incredibly complex. So too is the digital task.

The first step is building software able to recognise malicious traffic. One approach is to develop a program that creates digital 'antibodies' that attach to anything suspicious or unusual.

Suspicious files are automatically sent to a central location for analysis and scrutiny. For malicious traffic there are two outcomes – the virus signature is identified for future detection, and an antidote is created to counteract it. Then, the signature and antidote are spread around the network, ready for use when other computers are attacked.

engineers, programmers and mathematicians to work on cryptography, to write secure computer code and crime-fighting software, to create safer machines.

We need properly qualified citizens who can be security cleared and called on to help the Australian government. To this end the government should introduce a scholarship scheme to encourage a step change in the number of young Australians studying science, technology, engineering and mathematics.

And we need more women. In an increasingly digitised future we run the risk of seeing a professional and educational chasm re-open between men and women – who are already under-represented in this sector. Anecdotal evidence suggests girls, generally, need more persuading to engage in the three Cs. If we are to increase the number of women focusing on cyber technologies at tertiary institutions and in the workforce – bringing

a balance and skill set desperately required in the future – this must be addressed at primary and secondary levels.

Not only will such efforts lead to a safer and more secure Australia – and world – but an expanded Australian IT-security industry would be good for the economy in what is a fast growing multi-billion-dollar market. It makes dollars and sense.

ALASTAIR MACGIBBON specialises in internet fraud, consumer victimisation and internet security. He is managing partner of Surete Group, a director of the Centre for Internet Safety at the University of Canberra and CEO of the not-for-profit 'white hat' hacker-certification body, the Council of Registered Ethical Security Testers. Previously, he headed Trust & Safety at eBay Australia and later eBay Asia Pacific, was the founding director of the Australian High Tech Crime Centre, and a federal agent with the Australian Federal Police.

FURTHER READING

The ePayments Code covers electronic transactions, including ATM, EFTPOS, credit card transactions and internet banking.

MacGibbon, A. 2009, 'Cyber security: Threats and responses in the information age', Special Report 26, Australian Strategic Policy Institute, http://www.aspi.org.au/publications/special-report-issue-26-cyber-security-threats-and-responses-in-the-information-age.

Thomson, I. 2013, 'Researcher sets up illegal 420 000-node botnet for IPv4 internet map', *The Register*, http://www.theregister.co.uk/2013/03/19/carna_botnet_ipv4_internet_map/.

Smith, G. 2013, 'Microsoft takedown busts up global botnet cybercrime ring', *Huffington Post*, http://www.huffingtonpost.com/2013/02/06/microsoft-botnet_n_2632616.html.

Science, coupled with innovation, has
the potential to help develop solutions
to the world's challenges.

THE FRUITS OF SCIENCE

Innovation is a difficult journey from the lab to the real world, writes Craig Cormick

INNOVATION WORKS SOMETHING like this: a research scientist develops something brilliant, which is then developed into a product, is commercialised, the general public loves it and buys lots of it, so the developers become wealthy and the public find their lives greatly improved.

Sorry, we'll start that again.

Innovation works something like this: a research scientist develops something brilliant and goes through an extremely difficult process to develop it into a product, and after finally getting it to market the public are aghast/afraid/suspicious of the technology, so commercialisation fails and nobody is happy and few lives are improved.

To understand that the experience of turning brilliant science into successful and innovative products lies somewhere between these two scenarios, let's look at innovation before looking at the science.

There are numerous definitions of innovation. To save writing an essay, let's just call it doing clever stuff in a more clever way to get a good outcome. It can be a product, a process, a service. Its impact can be grand or incremental. And it can, like many things with multiple definitions, mean different things to different people.

For some, innovation offers a certain path to economic growth and social betterment; it leads to new industries or new goods and services, produced more efficiently, that people want to buy. What right-thinking person wouldn't love that, right?

In many instances this is true. Think smart phones, wi-fi, tablets, organic light-emitting-diode televisions, e-medicine, robotics. Brilliant science led to brilliant products that have huge consumer demand. Almost everyone is happier and better off.

But, of course, this isn't the way things always work out. That's a pity, of course, since planet-wide changes – growing population, expanding and shifting economy, changing climate and lifestyle expectations – create unique challenges to which we need new solutions. And science, coupled with innovation, has the potential to lead to the development of such solutions.

But only if we get the innovation thing right.

Europe accounts for about 75 per cent of the world's purchases in photovoltaics, their own manufacturing accounts for only about 3 per cent of it.

ADVANCE AUSTRALIA

It is becoming increasingly clear that the way Australia has traditionally managed innovation is unlikely to maintain national health, wealth and wellbeing in a rapidly changing global economy. Australia has dug up mineral resources and sold them overseas for many years, but that threatens to be less profitable in the future. Farmers have watched sheep graze and fatten up and then exported them. More than ever, agricultural production is subject to the vagaries of climate and competition. Meanwhile, we have seen the manufacturing sector contract, close down and move off-shore.

"There's no question that at some point our economy is going to have to shift and become substantially different from what it is now and be based on innovation," says Professor Ian Chubb, Australia's Chief Scientist.

There is a clear and growing chasm between where we are now and where we need to be. Clearly, the challenge for Australia is to find the best way to cross the chasm and move towards a sustainable economy that is less vulnerable than the one for which we must shrug off sentimental attachment, the one that provided the nation's prosperity to date.

Australia has certain strengths. We do good science. We are a creative nation – at times – and we know where we'd like to be. But we have a poor record of commercialising good science. We aren't always creative in the right areas and we have a poor understanding of what innovation is.

Unfortunately for the sentimentalists, 21st century innovation isn't based on the good fortunes of geography, geology and climate, as was much of our traditional wealth. Today, innovation only happens after careful planning, investment and strategic decision-making.

Even then the science ☐ technology ☐ commercialisation model doesn't always flow as smoothly as predicted. As an example, Europe is a strong performer in the science of photovoltaics. They publish papers at a high level and have a good patent track record. Still, they are not good at successful commercialisation of their research. Although Europe accounts for about 75 per cent of the world's purchases in photovoltaics, their own manufacturing accounts for only about 3 per cent of it.

This situation was described in 2012 by Christos Tokamanis of the European Commission. "A lot of R&D [research and development] dollars are being invested in Europe, but the question being asked is when we are going to see commercial returns?"

IDENTIFYING THE IMPEDIMENTS

So what are the main impediments to successful innovation? If the answer was simple, it would, of course, be similarly simple to remove those impediments. Instead, the impediments are complex and multiple.

The federal government's 2012 Innovation System Report cites lack of suitable education of management, lack of an innovative culture and an imbalance in government versus private spending on research and development, as reflected in the difficulty crossing the funding 'Valley of Death'.

The list goes on: lack of R&D growth in key areas; lack of business access to publicly funded research expertise; the large number of Australian patents taken up overseas rather than here; lack of mobility of researchers between academia and business; lack of harmonising intellectual property frameworks across the publicly funded research sector; and lack of a concerted national science, technology and innovation strategy.

There is also an increasing body of research highlighting the need to incorporate consumer needs into successful innovation strategies to ensure that once a new product or service is developed it will be accepted. Think of this as the 'Vale of Death': a product is developed that meets all the traditional innovation requirements, but does not align with consumer values or preferences.

PUBLIC ENGAGEMENT

Examples include cases where developers give the public a solution to a problem that most never saw as being their problem; for instance genetically modified (GM) crops as a solution to agricultural productivity. Members of the public may believe research has crossed an ethical divide that some people were unwilling to cross, as in, for example, embryonic stem cell research. Public rejection also occurs with solutions such as nanotechnologies, where the flow of information about the technology is dominated by misinformation on risk and concerns.

BACKING INNOVATION – FROM THE AUSTRALIAN INNOVATION SYSTEM REPORT 2012

Backing innovation is a proven strategy. Australian businesses that innovate are:

- 42 per cent more likely to report increased profitability.

- Three times more likely to export and 18 times more likely to increase the number of export markets targeted.

- Four times more likely to increase the range of goods or services offered.

- More than twice as likely to increase employment.

- Over three times more likely to increase training for employees.

- Over three times more likely to increase social contributions such as community enhancement projects.

And it's not just a matter of selling the products harder or trying to better explain the science behind them. Having spent a decade having discussions with anti-GM, anti-nanotechnology and anti-vaccination advocates, and even anti-climate change advocates, it is clear to me that their issues are rarely about the science.

Rather, it's about the personal values the science challenges. These can include issues relating to questions about messing with nature, monopolistic behaviour by agri-conglomerates or ethical fears over the misuse of genetic information. Aligning a product with people's values, through some form of public engagement to discover what those values are, will increase its chance of having a dream run. Clash with those values though and you could be in trouble.

Good early engagement allows the public to make choices about how they would like to see the research being developed. It also makes good marketing sense to ask the end-users what they want before it is produced. It makes poor sense to assume they will automatically share the developer's idea of what is needed.

For instance, based on what we have seen in terms of public rejection of large agri-chemical company involvement in GM crops, if the public had been consulted about GM science back in the mid-1990s, it's a strong bet they would have been against agricultural firms using it to develop broad-acre crops such as soya or corn that are resistant to their own pesticides or herbicides. It's much more likely they would have preferred to see non-food crops being used to produce pharmaceutical compounds, or healthier niche foods developed.

Some of the contentious innovative research currently in Australian laboratories includes artificial photosynthesis; affordable personal genome sequencing of unborn children; a universal flu vaccine based on mutated viruses; 3D printing of biological materials; and body organs and computer chip brain implants designed to boost memory or intelligence.

The potential benefit to Australian society from such projects is enormous. But for this they will need strong institutional support and community endorsement for their use and regulation, skilled developers to create them and sufficient funds to commercialise them.

These are big expectations. They will require a lot of very clever people working together in new ways to share old wisdom and new ways of thinking. Such next-generation teams must build a bridge that crosses the 'Prosperity Chasm', the 'Valley of Death' and the 'Vale of Death'. Building 80 per cent, or even 90 per cent, of a bridge won't be enough.

DR CRAIG CORMICK has managed community engagement programs for contentious technologies with the Innovation Department for more than a decade, and is widely published on issues relating to public engagement on science and technology. He currently works for the Commonwealth Scientific and Industrial Research Organisation (CSIRO).

FURTHER READING

Department of Industry, Innovation, Science, Research 2012, *Australian Innovation System Report 2012*, http://www.innovation.gov.au/innovation/policy/AustralianInnovationSystemReport/AISR2012/index.html.

Office of the Chief Scientist 2012, 'Breakthrough actions for innovation', http://www.chiefscientist.gov.au/2013/02/breakthrough-actions-for-innovation-released/ .

Cormick, C. 2012, 'Ten big questions on public engagement on science and technology', *DEMESCI: International Journal of Deliberative Mechanisms in Science,* 1(1).

Cormick, C. 2012, 'The complexity of public engagement', *Nature Nanotechnology,* 7, 77–78.

Cormick, C. 2009, 'Why do we need to know what the public thinks about nanotechnology?', *NanoEthics,* 3(2),167–173.

Cormick, C. 2005, 'Lies, deep fries, and statistics!!', *Choices*, 20(4).

SCIENCE DIPLOMACY

Engaging other nations in the pursuit of knowledge promotes international relations and national wellbeing, writes **Brendan Nelson**

The Australian Square Kilometre Array (SKA) Pathfinder telescope in WA is one of the precursor arrays for the planned multinational SKA – an example of how science can bring nations together.

THE JESUITS TAUGHT me that four virtues are essential to a 'successful' life. Unsurprisingly, perhaps, the first was compassion. Beyond the spiritual and humane, it's held me in good stead. In a practical sense, it has meant that I should try to understand how others think – to always ask myself how the person with whom I am dealing thinks and formulates his or her world view. Knowing what others think is one thing, understanding *how* they think is quite another.

Graeme Davison, in *The Use and Abuse of Australian History,* similarly observed that ethical and responsible citizenship relies on each of us being imbued with the imaginative capacity to see the world through the eyes of others.

As a non-professional diplomat arriving in Brussels early in 2010, it also seemed a sound basis for diplomacy. The European Union (EU), to which I was accredited, had just published its EU2020 vision. Its stated ambition was for a 'Smart, Sustainable and Inclusive Europe'. One of the seven ways of achieving this was to lift EU investment in research and development (R&D) to 3 per cent of Gross Domestic Product (GDP).

I have found in life that when fronting for a meeting, it's helpful to have something on the agenda that is also on the other party's agenda. As I thought about the 15 minutes I would

have, first with the president of the European Commission and then the president of the European Council, I asked myself how Australia might advance its own interests by also helping the Europeans meet theirs.

Research co-operation seemed an obvious area for discussion.

At the time, we were also contemplating our bid for the next-generation radio telescope, the Square Kilometre Array (SKA). As Australia's relationship with the EU had historically been framed by us narrowly around trade conflict in agriculture and market access, I was anxious to broaden the engagement into different areas. Simply saying to the bloc's leaders that Australia should be awarded the SKA on the basis of its scientific research quality, seemed likely to have negligible impact.

We needed to position ourselves as being helpful to Europe in the field of science and, in doing so, demonstrate the strength of Australia's research output.

So, when meeting President Barroso to present credentials to the European Commission, I went with four issues. He listened carefully to the first three with the polite, interested air of a seasoned politician with a busy day ahead.

Then I got to science.

I told him I had noted the R&D ambition set out in the EU2020 agenda. While agreeing that was one of the key levers the European nation states could pull to lift themselves from the economic crisis engulfing them, I suggested their target was unlikely to be achieved. A number of finance ministers were already publicly stating they could not – and would not – invest more money in research.

I went on to explain that Australia produced 3 per cent of the world's intellectual output in publications. In some disciplines it's much higher. I reminded him of our Nobel Prize winners, members of the Royal Society and citation rates, among other things. I told him we realised, however, that to go further, we needed to lift our levels of international collaboration substantially. Further to this, if the EU was to get anywhere near its 3 per cent target, one key device for doing so would also be collaboration.

I had his attention.

I proposed that Australia and the EU could pool some of our research money in support of joint projects. This could be at both the researcher and institutional level. The specific projects to be funded would be selected on a peer-reviewed basis by our respective scientific communities. But together we could nominate areas for joint research collaboration. I proposed water-resource management, agricultural productivity, energy efficiency, renewable energy and human adaptation to new and emerging technologies by way of examples.

I pointed out to President Barroso that if such a model was developed and implemented it would not only deepen co-operation between my country and the EU, it would leverage up our respective research spend. I told him that while Australia's research budget was small in comparison, such a model applied to similar like-minded countries had the potential to be a 'win-win'.

He got it, telling me with some enthusiasm that he saw considerable potential in this idea. He said I must see the EU commissioner for research, Máire Geoghegan-Quinn.

Given that EU nations produce 38 per cent of the world's scientific papers and the European

Neither Australia nor the US has fulfilled their respective, considerable potential in science diplomacy. We should – be it with the Islamic world or the Asia Pacific. Our future depends on it.

Commission has a budget for research and innovation exceeding 120 billion euros (about AU$170 billion), the potential for Australia was, and remains, significant. Collaborations between Australian and EU scientists exceed those with the US.

When I debriefed staff at the Australian embassy on my meeting and discussion, Dr Martin Gallagher, our then indefatigable science counsellor said, "Brendan, excellent science diplomacy". I presumed this was an expression thrown around in the science community as part of the jargon used in his world. It was not something I had heard before, even though as science minister in the last decade I had discussed scientific co-operation with my counterparts in a number of countries.

I have since discovered that is far from the case.

Then Prime Minister Gillard came to Brussels some months later and similarly raised and discussed research co-operation with her two key EU interlocutors. The conversation continued when Barroso visited Australia the following year. It was certainly a change from the traditional confrontational nature of dialogue we had experienced in the past, lambasting the Europeans for agriculture and trade policies.

From this early, positive engagement we were able to build a relationship with the EU that was much more constructive. The multiplier effect for increased co-operation in education was an early, positive impact. When the prime minister stood next to the president of the EU in October 2010 and said that we would be working towards closer co-operation in research, its head-turning effect was not confined to the research community. In education, the EU's education commissioner and director general both expressed an interest

Australian Research Connections

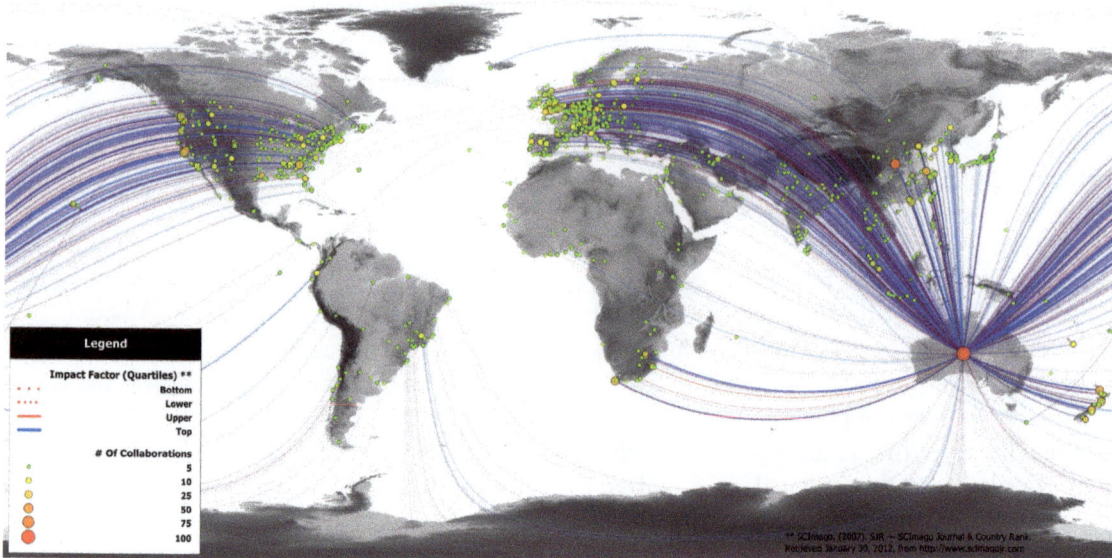

The global nature of science shown by the frequency and location of Australian research papers in environmental science co-authored with international partners in 2010.

in exploring further co-operation and linkage of programs. I even found myself invited as ambassador to a small meeting of 10 to read and comment upon the Commission's draft directive (white paper) for higher education.

So too, as the EU established its own diplomatic service with the European External Action Service, Australia's relationship reached further into foreign affairs and security.

Similarly, in overseas development assistance, environment, climate change, economic issues and the G20 agenda, I endeavoured to shape our overarching engagement around areas where we might advance a common interest. That interest, wherever possible, would be evidence- and research-based.

I've since learnt that science diplomacy is the use of scientific collaboration between and among nations that seeks to meet common challenges. If skilfully applied, it becomes the foundation for constructive international partnerships.

The Obama administration has been particularly enamoured with Joseph Nye's 'Soft Power', as described in his 1990 book, *Bound to Lead: The Changing Nature of*

American Power. Greater investments by the US in areas such as science diplomacy are regarded as essential to complement its economic and military power.

President Obama's 2009 'New Beginning' speech in Cairo, in which he sought to reach out to the Muslim world, argued for a greater engagement by the US through science, technology and innovation. He saw the linking of scientists as a basis for common understanding.

The world has undergone a major transformation in less than a decade, the scale of which is not well appreciated by most Australians. The shifting of political and economic power from North America and Europe to the Asia-Pacific over the past 20 years was rapidly accelerated by the Global Financial Crisis and the world's response to it.

China's re-emergence and the critical shape of the template for its relationship with the US are critical to our future. As Henry Kissinger observed, the template for that relationship is being forged now in our region. It is into this global and regional environment that science diplomacy can and should play a pivotal role.

Australia and China have had to deal with a number of 'irritants' in the past five years. But there is resilience in the relationship now that gives each country increased confidence in extending its depth and breadth. China's challenges are immense, among them the need to lift 150 million people out of poverty, create 24 million jobs a year, provide housing to 10 million still homeless, energy security, environmental sustainability and the urgent need to address urban pollution.

Australia has much to offer Chinese researchers and its scientific leaders in expertise relevant to addressing these and other challenges. Irrespective of the diplomatic, trade and economic tensions that ebb and flow between our two countries, China's political and intellectual leadership can only appreciate research collaboration that helps lift understanding while addressing real issues bearing down on the country. Again, it is imperative that we actively pursue areas of common interest and concern in science as one means of adding substance and ballast to the relationship. Australian researchers actively working with their Chinese counterparts on agricultural productivity and clean energy technology alone will serve our mutual interests.

As China rapidly boosts its investment in research and relentlessly drives the transformation of a handful of its leading universities, the scientists within them will largely shape the direction in which their nation goes. Its challenges – economic, environmental, agricultural, health, social and diplomatic – will all require guidance from science. Researchers in humanities and social sciences will be no less important in this regard.

Australia has skilfully managed its relationship with China, balancing its economic needs with concern for human rights and rule of law. We have a program for research co-operation, but there is much more we could do to deepen the ties between men and women whose passion is knowledge. Similarly, those Australian companies that join our political leaders in their visits to countries such as China are themselves, to varying degrees, conduits for science diplomacy where new knowledge is the basis for wealth creation.

While the East Asia Summit, now including the US and Russia, traverses security issues, its agenda is largely in areas where Australian scientists, working with their counterparts, can collaboratively inform reasoned decision-making in key areas ranging from climate change to food production.

Neither Australia nor the US has fulfilled their respective, considerable potential in science diplomacy. We should. Our future depends on it; whether it's the West reaching out to the Islamic world or deepening stable multilateral integration in the Asia Pacific.

The other three Jesuit values for success – commitment, conscience and courage – will all need to be applied to this task. It also requires sustained, whole-of-government political will and resources.

DR BRENDAN NELSON is director of the Australian War Memorial. He served as Australia's ambassador to the European Union and NATO for three years from 2010 and as Minister for Education, Science and Training from 2001 to 2006.

FURTHER READING

The Royal Society 2010, 'Miliband urges greater role for science in diplomacy', http://royalsociety.org/news/science-diplomacy/.

Armitage, R.L. and Nye, J.S. 2007, 'Stop getting mad, America. Get smart', *The Washington Post*, http://www.washingtonpost.com/wp-dyn/content/article/2007/12/05/AR2007120502254.html.

Dickson, D. 2010, 'Can science diplomacy help strengthen the Muslim world?', *SciDev.Net*, http://scidevnet.wordpress.com/2010/06/26/can-science-diplomacy-unite-the-muslim-world/.

The Royal Society 2010, 'New frontiers in science diplomacy', http://royalsociety.org/policy/publications/2010/new-frontiers-science-diplomacy/.

POWERING THE FUTURE

The clock is ticking on the drive for sustainable energy, writes **Barry Brook**

ACCESS TO CHEAP and reliable energy has underpinned Australia's development for decades. Fossil fuels – coal, oil and natural gas – provided the concentrated energy sources required to build our infrastructural, industrial and service enterprises. Yet it's now clear this dependence on carbon-intensive fuels was a Faustian bargain and the devil's due, because the long-run environmental and health costs of fossil fuels seem likely to outweigh the short-term benefits.

In the coming decades, Australia must tackle the threats of dangerous climate change and future bottlenecks in conventional liquid-fuel supply, while also meeting people's aspirations for ongoing increases in quality of life – all without compromising long-term environmental sustainability and economic prosperity. Fortunately, there are science and technology innovations that we could leverage to meet these goals.

SEEKING COMPETITIVE ALTERNATIVES TO COAL

How can Australia shift away from coal dependence and transition to competitive, low-carbon alternatives, and what role will science and engineering play in making it happen? To answer these questions, a key focus must be on electricity-generation technologies – electricity is a particularly convenient and flexible 'energy carrier' – and to consider the key risks and advantages with the alternative energy sources that will compete with fossil-fuel power.

In 2012, the majority of Australia's electricity was generated by burning black and brown coal (75 per cent), with smaller contributions from natural gas (13 per cent), hydroelectric dams (8 per cent) and other renewables (4 per cent). The nation's installed capacity now totals more than 50 gigawatts of power generation potential, with electricity and industrial-heat energy production currently resulting in the annual release of 285 million tonnes of carbon dioxide, about 52 per cent

Variable renewables, such as solar collectors, are 'clean' in that they harness energy that is constantly being replenished.

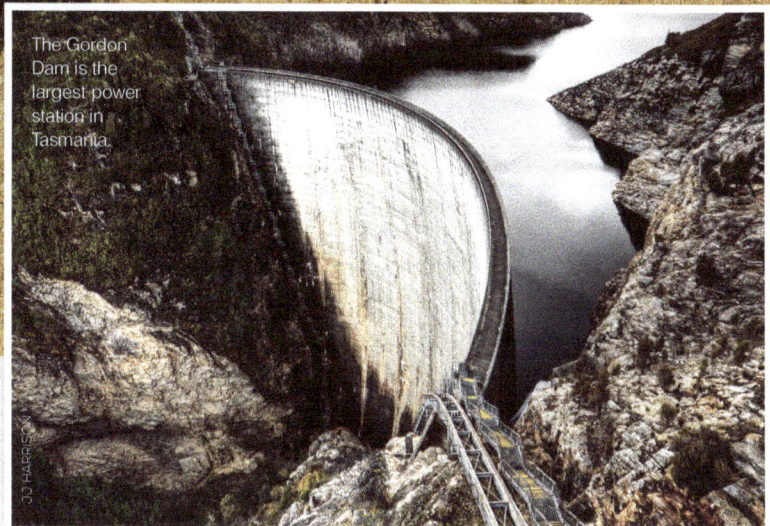
The Gordon Dam is the largest power station in Tasmania.

of our total emissions (see: **Understanding electricity units**).

Clearly, the non-electric energy-replacement problem for Australia would also need to consider transportation and agricultural fuel demands. In a world beyond oil for liquid fuels, we will need to eventually 'electrify' most operations: using batteries, using heat from power plants to manufacture hydrogen from water, and by deriving synthetic fuels such as ammonia or methanol.

Under 'business as usual' forecasts produced by government energy analysts, electricity use in Australia is expected to grow by 60 to 100 per cent through to 2050, with hundreds of billions of dollars of investment needed in generation and transmission infrastructure just to keep pace with escalating demand and to replace old, worn-out power plants and transmission infrastructure. At the same time carbon dioxide emissions must be cut by 80 per cent to mitigate climate-change impacts, via some combination of enhanced energy conservation and new supply from clean energy sources.

AN UNCERTAIN MIX OF FUTURE OPTIONS

Although there are a huge number of *potential* energy options now being developed that *might* one day replace coal in Australia, not all alternatives are equally likely.

The government renewable-energy target has Australia aiming to derive about 30 terawatt hours of electricity from some combination of wind, solar, wave and geothermal (hot rock) energy by 2020, compared to 10 terawatt hours

UNDERSTANDING ELECTRICITY UNITS

Dynamic electricity is the motion of electrons, and can be used to 'do work' (e.g. cause lights to shine, turn electric motors, run televisions or air-conditioner compressors). Electrical current is constantly pushed out of power stations through transmission wires, to your home or business. Power is the rate of flow of electricity, and is measured in watts. A solar panel on your roof might produce 1500 watts, whereas a large coal plant produces one billion (a gigawatt). Energy is the amount of power delivered over a period of time. So if that rooftop solar panel operated in full sun for one hour, it would produce 1500 watt-hours (1.5 kWh) – enough energy to boil a typical kitchen kettle about 10 times. A coal plant running continuously for a year will produce an enormous amount of electrical energy – about 8760 gigawatt-hours, or 8.76 terawatt-hours (TWh). Australia consumed 227 TWh in 2010. People pay for electricity in cents per kWh.

from these non-hydro renewable sources in 2011 (see: **Understanding electricity units**). The target is driven by legal mandates and financial incentives. But beyond this modest market penetration their future is uncertain.

The challenge is daunting. Although there are promising low-carbon exemplars – Norway gets almost all its electricity from hydro, and France derives 80 per cent from nuclear energy – historically, no country has achieved a penetration of solar or wind beyond about 20 per cent of supply. Denmark has the highest penetration of new renewables, per capita, based largely on wind power. Yet even the Danes still rely on domestic coal and imported nuclear and hydro to meet their reliable baseload power needs, and have high domestic electricity prices.

If Australia is to push significantly beyond the 20 per cent for non-hydro renewables to 50 per cent or more by 2050, this expansion will need to be underpinned by stunning advancements in large-scale energy storage with significant cost reductions, along with ubiquitous 'smart grid' technologies (see: **Energy efficiency** on p98

for more details) to balance supply-demand and improve the efficiency with which energy is managed. All this is possible but the key issue of whether variable renewables with storage can win the cost-benefit competition remains a large unknown.

FIT-FOR-SERVICE CLEAN ENERGY

To achieve the long-term goal of phasing out coal and slashing greenhouse-gas emissions, Australia must actively invest in the science, technology and commercial demonstration of next-generation electricity infrastructure. Ideally, the underpinning technologies will be fit-for-service, low-carbon 'plug-in' alternatives to fossil energy that are scalable, reliable and cost-effective, while also balancing issues of societal acceptance and fiscal and political inertia. Also, although infrastructure and fuel costs will be critical considerations, a technology must also

The wholesale replacement of fossil fuels with cleaner alternatives represents a massive global conundrum, because we have to continue to provide reliable and affordable energy while reducing greenhouse-gas emissions.

constitute 'value for money' by delivering its intended service adequately.

For a new electricity generator to serve as a direct fit-for-service replacement for coal, it should be dispatchable (i.e. can be delivered on demand), without need for large or expensive external storage, or else have a reliable fuel supply. It must also have low or moderate carbon-emissions-intensity, and be able to produce at a high capacity factor (i.e. delivering a near constant supply across a 24-hour period or longer). This can potentially be achieved in diverse ways, via a portfolio of non-carbon or low-carbon fuels (nuclear and biomass), the tapping of constant or predictable flows of natural energy from geothermal, tidal and wave power, and by geographically spreading wind and solar collectors across nationwide grids, from urban zones to farmland to deserts.

Clearly, different electricity technologies will fill a variety of niche roles in future markets. Variable renewables, such as wind turbines and solar collectors, are 'clean' in that they harness energy that is constantly being replenished, and they enjoy strong community support. Their use will continue to grow in Australia, especially as deployment costs are reduced.

Three new and promising substitutes for coal that Australia might consider seriously pursuing are advanced nuclear reactors based on small modular (factory-built) designs, deep-earth 'hot dry rock' geothermal, and ubiquitous small-scale solar. These clean-energy technologies offer many attractive and often complementary features, yet none is currently cost-competitive with coal, oil or gas, being technically immature or at least unproven at a large scale. As such, all are considered financially risky. There are real and exciting opportunities for science to work to improve and demonstrate these innovative new technologies, but technologists must collaborate with government and industry to ensure the markets are ready.

Ultimately, there is no magic-bullet energy source that can solve all problems perfectly without any negative impacts on society and the environment (see: **No perfect energy technology**). Given this reality, energy plans that expand the role of both nuclear and various promising renewable and energy-storage technologies, and allow them to compete on a level playing field, seem to make most sense.

'BIG SCIENCE' FOR INNOVATIVE ENERGY FUTURES

The wholesale replacement of fossil fuels with cleaner alternatives represents a massive global conundrum, because we have to continue to provide reliable and affordable energy while

This wind turbine under construction, the E-126 model, is 200 metres tall and can generate 7.5 megawatts of electricity at peak output.

reducing greenhouse-gas emissions. So what can Australia – a relatively small part of the world economy – do most effectively to contribute to this effort, beyond supporting basic and applied scientific research?

Australia has been a world leader in the development of lower-cost and more-efficient crystalline solar photovoltaics, and is working at the scientific cutting edge in the research and development of new organic solar cells and solar-thermal dish technology. Support for such work should definitely continue, but we must not shirk from some risk taking, by engaging in more controversial fields like next-generation nuclear and carbon-capture-storage.

NO PERFECT ENERGY TECHNOLOGY

Fossil fuels have provided modern society with a cheap and convenient source of concentrated, stored energy. This has allowed us to build our industrial economies, information technology, global food production and transport systems, and many other foundations of society. However, people are also well aware of the damage they cause to the environment – explosions, chronic health issues from soot and heavy metals, and greenhouse gases.

Nuclear fission, an abundant and low-carbon energy source, has an enormous and proven potential to supply reliable baseload electricity and displace coal or gas power plants directly. Yet the prospect of nuclear energy concerns many people who worry about sustainability, spent-fuel disposal and radiation release from accidents. Innovative new designs like the integral fast reactor and liquid fluoride thorium reactor technologies (see: 311 megawatt PRISM module, right) could, if commercialised, avoid or heavily mitigate these hazards, by incorporating passive safety with inherent self-protection, and by recycling nuclear waste to generate zero-carbon electricity. However, no reactor can be made perfectly safe, and so, as is the case with technologies such as cars, planes, food supply, electricity and medicine, society must tolerate some level of risk.

The 311 megawatt PRISM module is a next-generation nuclear power plant. Passively safe and able to recycle its waste, it represents a potential clean-energy frontier for fission-based electricity technology.

Clean-energy alternatives to nuclear are not without their hazards, on top of the concerns of on-demand reliability. The life-cycle greenhouse-gas emissions of photovoltaics are higher than nuclear power, and manufacture of solar cells uses a mix of highly toxic chemicals that do not degrade over time. In the United Kingdom there were 1500 wind-power-related accidents and four fatalities between 2007 and 2011. Wind turbines and solar-thermal plants use large quantities of concrete, steel and land per unit of electricity delivered, compared to nuclear or geothermal alternatives. Hydro requires massive land transformation and intermittent renewable energy sources typically rely on natural-gas back-up. There are also many 'system factors' to manage, such as intermittency-related effects on the stability and congestion of transmission networks.

Such problems do not mean that large-scale renewables are not worth pursuing or that advanced nuclear is the only viable option. But it does emphasise the fact that we must avoid arbitrarily closing off technology options without looking at the big picture.

All of the above are vital components of a cost-benefit analysis. Trade-offs are inevitable – there is no ideal energy option and urgency will help dictate the response. Ultimately, science can provide the range of potential options, but society must make the final choices.

Above all, we must ensure that scientifically grounded, government-led climate strategies are clearly focused on measurable and timely reductions in greenhouse-gas emissions, with the rapid deployment of high-capacity, clean-energy technologies that work as direct substitutes for fossil fuels. It will mean strategic investment in a portfolio of potential 'winners', while accepting that there will probably be more failures than successes.

In this context, it's necessary that the best technologies get their due opportunity by taking a firm, hands-on approach. Time is not on our side. This will require a sustained and timely injection of money from a coalition of nations to create the manufacturing, distribution, education, security and skills base that is absolutely necessary for a 21st century re-imagining of energy.

I'd argue that Australian participation in flagship big-science projects, based on active

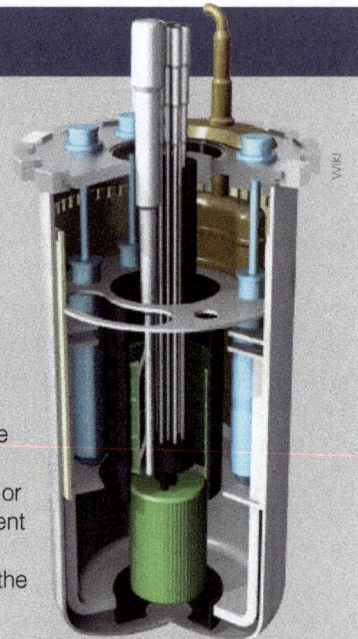

Several older black coal-fired power stations are being sold off as society transitions to new energy sources.

multilateral engagement, will be crucial to making deep inroads and reducing costs. Examples such as the Large Hadron Collider and Square Kilometre Array are motivating illustrations of how such endeavours can work. In the energy realm, the $20 billion International Thermonuclear Experimental Reactor project for future fusion power is an archetype. But similar collaborations are needed for a range of the most promising technologies, with Australia contributing to both innovative science and ongoing financial support.

To realise this vision, Australia also needs an educated and engaged society that understands and embraces the need to pursue new – often unfamiliar – energy technologies. This will require a strong and sustained focus in the school curriculum on science and engineering, coupled with improved public outreach for all ages. People must be encouraged to think critically about sustainable energy options and the trade-offs Australia must inevitably face.

PROFESSOR BARRY BROOK is an environmental scientist who models global change and holds the Sir Hubert Wilkins Chair of Climate Change at the University of Adelaide.

FURTHER READING

CSIRO eFuture – explore scenarios of Australia's electricity future. http://efuture.csiro.au/.

MacKay, D.J.C. 2013, *Sustainable Energy – Without the Hot Air*, UIT Cambridge, Cambridge.

Science Council for Global Initiatives 2012, 'The case for near-term commercial demonstration of the integral fast reactor', white paper, http://www.thesciencecouncil.com/pdfs/whitePaper.pdf.

ENERGY EFFICIENCY

Energy efficiency has the potential to solve many of the energy challenges facing Australia, and exciting science is needed to realise this potential, writes Glenn Platt

WHAT'S THE FIRST thing you think about when you read 'energy efficiency'? Cold showers? Warm beer? Perhaps those lab-coated energy cops telling you to turn off the TV at the wall. It's hardly compelling stuff – as one of my colleagues once lamented, "It's just not sexy!"

It might never be sexy, but energy efficiency research is at the cutting edge of science, way beyond the compact fluorescent globes or insulation batts that most people associate with the field.

While addressing challenges such as the cost of electricity or where our energy is supplied from, this work is also improving our understanding of how people make decisions, develop new materials, and design computer systems that work by specifically targeting some of those unique characteristics that make us human.

At its heart, energy efficiency is about doing the same thing, but with less energy. It is about industries as well as individuals being less wasteful, while maintaining productivity and current lifestyles. The level of wasted energy involved in day-to-day activities would surprise many of us. Typically, coal-fired power stations convert just 30 per cent of the energy in coal to generate electricity – 70 per cent of the energy in that coal is wasted. Of this 30 per cent we keep, up to 10 per cent is then lost in distribution and transmission. Considering a typical electricity load such as the halogen downlights found in many Australian homes, these convert just 3 per cent of the energy supplied from that power station to light.

This means that any energy saved at the end of the supply chain – in our offices and homes – translates to approximately a fourfold saving in energy at the start of the chain, at the power station. This magnification effect illustrates just how important energy efficiency is: if we can just use one less unit of energy at the end of the supply chain, we'll need far less at the start.

Science is driving an incredible range of exciting technologies that will change the way we use energy. One major area of focus in energy efficiency is on air-conditioning and heating systems, which make up the majority of the energy use in many Australian buildings.

IN WITH THE NEW

One of the most exciting areas of energy-efficiency research is the design of completely new types of air-conditioning systems.

Air conditioning uses massive amounts of energy, and leads to other challenges such as strained electricity networks during particularly hot days of the year. While the energy efficiency of air conditioners has steadily improved, the changes have been relatively slow, and the way air conditioners operate today remains very similar to when they were invented 100 years ago.

Today, scientists at CSIRO and the Australian National University are developing an entirely new type of air conditioner, one that challenges logic. This type uses the heat from the sun to cool our buildings. In this way, solar heat replaces electricity as the source of energy that drives the cooling process and, importantly, the hotter the day, the better the air conditioner works.

One such system uses evaporative cooling, which cools air by evaporating water, a technology that has been around for decades. Evaporative coolers work well in dry conditions, but unfortunately do not work well in humid coastal environments where most Australians live. To solve this, Australian scientists from CSIRO's Division of Materials Science and Engineering are developing new materials that first absorb the moisture from the air inside the building, removing the humidity so it can be cooled by an evaporative cooler.

These scientists are textiles experts – several years ago they were developing ways to process wool to get the comfortable merino clothing we wear today. Now, they are developing new fabrics that sit inside an air conditioner, absorbing moisture. Together with the engineers combining these components into a working system, these scientists will dramatically change the type of air conditioner we find in our buildings.

IMPROVING THE OLD

While completely new types of air-conditioning systems may be great for new buildings or system replacements, this technology won't help all those buildings out there that already have a conventional air conditioner.

For these buildings, scientists are using state-of-the-art computer and 'human factors' science to design new ways to control existing air-conditioning systems to save a lot of energy. For example, traditional air conditioners work by keeping the temperature inside quite constant – usually about 22°C, all year round. It turns

The solar tower and field at the National Solar Energy Centre converts solar energy into eletricity.

out that humans don't need such a constant temperature to be comfortable – if it's 38°C outside, most people will find 25°C inside perfectly comfortable, and keeping it 25°C inside uses much less energy than cooling to 22°C.

Using modern computer science, and research into what makes humans comfortable, scientists are developing control systems that allow the air conditioners to adapt to the temperature outside, to human comfort, and even to changing conditions such as trees shading outside walls during parts of the year. These new controllers are essentially small computers that can be added to existing air-conditioning systems, helping them save huge quantities of energy, while improving people's comfort in the building.

BEHAVIOURAL CHANGE

While some energy-efficiency scientists are working on new devices to save energy, others are working on a completely different challenge – people. These scientists are trying to figure out why people make the energy decisions they do, and what information may help people make better, more efficient-energy decisions. These scientists, from backgrounds such as psychology or social science, are researching people's behaviour.

Traditionally, energy usage has largely been driven by unconscious, automatic and habitual behaviours, and is not typically something people often think about. This situation is changing. Because of issues such as rising costs or concern about energy supply and climate change, people are paying more attention to their energy consumption, and there is an opportunity to change traditional behaviours. The challenge is – what information should be provided to people to encourage more efficient choices? We are confident people can be just as comfortable, or have just as much fun, using the highly efficient technologies available today. But how do we encourage users to choose these alternatives?

When considering how to motivate behaviour change, researchers are looking at the messages presented to consumers about energy. Consider two very different messages with a similar goal: "If you reduce your energy consumption, you will save $50 on your electricity bill", or "If you continue to consume energy as you are, your electricity bill will increase by $50". These messages say almost the same thing, yet the first one is seen as a gain, and the second as a loss. Another alternative could be: "Your consumption was much more than others in your street", or even, "By reducing your energy consumption, you will help reduce our impact on the climate our grandchildren will experience".

As can be seen, there are many approaches to the same messaging campaign, yet surprisingly little is known about which approaches work best, particularly in an energy-efficiency context. Australian scientists are improving our knowledge here, trying to understand why people make the decisions they do. Ultimately, if people can be helped to make better decisions, huge energy savings can be realised – including those from the new energy-efficiency technologies other scientists are developing.

Interestingly, even when the science starts to take hold across society – when our air-conditioning systems are really efficient – or people are making better energy decisions, the challenges for scientists will continue. Currently most of the energy consumed over the lifetime of building is used by the occupants of the building, e.g. for heating and cooling. Eventually, that will change: over the life of a building, more energy will go into making the

SMART METERS

One energy technology getting a lot of attention at the moment is smart meters, a new type of electricity meter. A smart meter is simply a device that allows us to get much more information about our electricity usage, more quickly than before. Rather than getting one bill every three months and wondering where you spent that energy, a smart meter is like a speedometer in a car – it allows us to see how much electricity we're using, right now. Having better and more timely information about how we are using our electricity can only be a good thing – suddenly, as soon as we turn on the air conditioner, we'll see exactly how much it is costing us to operate. Unfortunately, at the same time as deploying smart meters into homes, some electricity companies changed their electricity tariffs, so some people associate a smart meter with a bigger electricity bill. This need not be the case; ultimately a smart meter is just a way of quickly reading and reporting consumption. How that information is used in bills isn't the meter's fault!

building than keeping the people inside the building cool, warm, and well lit once it is built. This will be a profound shift and mean a totally new challenge for Australia's scientists: how to make buildings that don't need much energy to run, keep people comfortable, and only need a small amount of energy in their materials and construction.

DR GLENN PLATT leads CSIRO's work in 'local energy systems', which includes energy efficiency, and how to maximise the benefits from renewable energy generation. He has a PhD in electrical engineering.

FURTHER READING

Solar cooling – CSIRO is developing new technologies and systems that use solar thermal energy to provide low-emission cooling – or heating – for buildings and refrigeration for food storage.

Sullivan, D., Armel, C., Todd, A. 2012, 'When "not losing" is better than "winning": Using behavioural science to drive customer investment in energy efficiency', American Council for an Energy-Efficient Economy Summer Study.

Laskey, A. 2013, 'How behavioural science can lower your energy bill', TED Talk, http://www.ted.com/talks/alex_laskey_how_behavioral_science_can_lower_your_energy_bill.html.

FUTURE ELECTRICITY SYSTEMS

Homes and suburbs are set to become the 'virtual power stations' of tomorrow, writes **Mark Paterson**

THE TRANSFORMATION

of Australia's electricity system has begun. With global energy challenges increasing the process will gather pace, becoming a wholesale redesign of the approach to electricity.

Over the next few decades Australia's electricity grid (see: **Key terms**) is likely to be transformed from the centralised, fossil-fuel-driven power delivery systems of today into increasingly decentralised, low-carbon, 'electricity transaction systems' where customers have almost infinite choice. The home, business and suburb will become core elements of an efficient new energy system, or 'smart grid'. As Michael Valocchi of IBM said in 2009, "The world's electricity network will change more in the next 20 years than it has in the last 100".

That sounds great. So why not "just do it"? For starters, contemporary electricity grids, which have been referred to as the "the world's largest machine", have changed little since World War II. Mid-20th century technology is clearly inadequate for the needs of a resource-constrained 21st century. However, redesigning grids for different purposes and greater efficiencies is a monumental challenge for any nation. And in the case of developed nations like Australia it's a bit like rebuilding the plane while flying it.

While change will be difficult, the need for it is becoming acute – particularly in the residential areas of Australian cities and towns – primarily for the following two reasons.

AUSTRALIANS' LOVE OF AIR CONDITIONING

First, Australian households adopted air conditioning *en masse* over the past decade. They generally only run them during short but extreme heat waves or cold snaps. Still, when operating, air conditioning uses a lot of electricity so the 'poles and wires' infrastructure running down each street has had to be oversized to service less than 100 hours' use per year. This

is akin to building a 10-lane freeway to the coast to prevent traffic jams on the Easter weekend.

Until recently, there were no alternatives to the very expensive approach of building grid capacity. While media outlets and politicians sometimes refer to this as 'gold plating' no one would be happy if the lights went out every time there was a cold snap. However, this is hardly an efficient way to use national resources, especially when electricity bills have to increase to cover the cost.

Worse, people without air conditioning pay more as a result. This is because the way electricity is billed traditionally focuses only on total electricity consumption (or volume) each month or quarter, measured in kilowatt-hours (kWh). This provides no bill reduction for those households that put little pressure on the electricity system because their electricity use is quite stable.

This is unfortunate because the costs of servicing air-conditioned residences that experience short but extreme spikes in electrical demand are not fairly apportioned to those households. Such spikes from thousands of air-conditioned homes at the same time – for example, at 5pm on a hot summer weekday afternoon often force the expansion of the grid capacity. The recent electricity bill increases experienced by all Australians have been heavily influenced by the need for this extra grid capacity, which is very expensive to build.

SOLAR ENERGY AS A 'SAFE HAVEN' FROM ELECTRICITY PRICES

Second, in response to the rising price of electricity, another trend is emerging across a growing proportion of homes in many developed economies, including Australia. Recently, the one millionth Australian home was fitted with solar photovoltaic (PV) panels. This represents the gradual transition of many households from being passive electricity consumers (receivers) to becoming

Australian households adopted air conditioning *en masse* over the past decade.

'prosumers' who both supply and consume electricity (producer-consumers).

At face value, the trend to renewable sources such as solar energy is encouraging, but there is an unintended downside. Unknown to most Australians, the mass adoption of solar photovoltaic generation (solar PV) has influenced the price of grid-supplied electricity. That's because the electricity system – including the way power companies charge for electricity – was designed for electricity flowing in one direction, from the generator to the consumer.

In other words, the multi-billion dollar electricity grid was designed and built at a time when it was inconceivable that most consumers would have on-site generation capable of pushing electricity back into the grid, in the opposite direction of normal flow.

Moreover, while a stable and reliable supply of electricity is taken for granted, the output from solar PV is entirely dependent upon whether the sun is shining. Cloud movements can cause the supply of electricity to fluctuate dramatically, and the output of

solar PV is negligible during extended periods of rain. As a result the electricity grid must provide back-up electricity while also requiring expensive new technology to cope with the rapid fluctuations of output from home-based systems.

Finally, individual solar PV systems have not traditionally been well integrated with the electricity system to ensure their output provides value to the wider community. While traditional centralised generation must be ramped up and down to match the community's demand for electricity at any moment of time, individual solar PV installations have no such mechanism.

Not surprisingly, inefficiencies arise from breaching the basic economic principle of matching supply of a commodity to demand for it. For instance, at midday on weekdays in dormitory suburbs it's common for solar PV output to exceed local consumer demand as most residents are away at work or school. The undesirable results can be twofold: grid voltage levels exceeding operating limits and

Recently, the one millionth Australian home was fitted with solar photovoltaic panels. This represents the gradual transition of many households from being passive electricity consumers to producer-consumers.

the lifespan of local household appliances potentially being shortened.

Taken together, these factors increase the price of the electricity supplied by the grid at night and during long periods of rain. Wind power similarly only provides electricity when the breeze blows. Again, because of the way electricity is traditionally billed, Australians who don't have or can't afford solar PV are significantly impacted by such price increases.

One solution could be for households with solar PV to install large battery banks and disconnect entirely from the grid. However, as large battery storage remains very expensive, the cost of total disconnection is likely to remain prohibitive in the medium term for all but the very wealthiest of households. Price estimates in 2013, for example, range from $20 000 to $80 000, depending on home size, construction type, occupancy level and lifestyle factors. In real terms, this would translate into actual electricity prices of more than triple the existing kWh tariffs.

A PROBLEM OR AN OPPORTUNITY?

Australia – and the world – needs low-carbon, reliable and affordable electricity systems. Doing two of these things is relatively easy. Doing all three at once is an enormous challenge.

The traditional approach to electricity is both reliable and relatively low cost, but not low carbon. As the connection of low-carbon generation such as solar or wind increases, the greater the need to incorporate expensive grid management and energy-storage technologies which can push up overall electricity prices. And, while wind and solar power certainly

provide low-carbon electricity, as stand-alone technologies they are less reliable due to their dependence upon variable weather patterns. They can also be expensive because of their need for supporting technologies such as energy storage.

Regardless, it's likely that by 2050 nearly 50 per cent of Australia's electricity will be generated locally: at homes, businesses, shopping centres and industrial complexes. Consequently, ensuring reliable, low-carbon and economically efficient outcomes will require homes, businesses and suburbs to become core elements of a well-integrated new electricity system. Rather than being considered as disconnected parts, this new approach will function more like an 'ecosystem' of interdependent parties that adapt to serve each other's needs.

As noted previously, both demand for and supply of electricity are highly variable in a low-carbon energy system. To make such systems affordable it's important to reduce the amount of expensive battery storage required to fill the gaps between lulls in supply and peaks in demand, and vice versa.

One great way of doing this is to create a system where thousands of electrical loads can be dynamically ramped up and down to match the actual generation output of solar PV (and wind farms) at any time. While this

The traditional electricity grid is characterised by a one-way flow of electricity from distant power stations delivered via high voltage transmission lines to residential, commercial and industrial consumers.

is something many countries are aspiring to do, Australia has a distinct advantage because we lead the world in standards development that enable many thousands of household appliances to participate in such a system, most notably the Australian Standard AS/NZS4755 series – "Demand response capabilities and supporting technologies for electrical products" – which applies to remotely controlled demand management of electrical systems.

THE HOME OF THE (NEAR) FUTURE

In practice, homes in a low-carbon future must function as an important part of the wider energy 'ecosystem'. This is why CSIRO commenced its Virtual Power Station (VPS) research several years ago, before many of these challenges were widely anticipated in the general community.

Building upon this work, CSIRO's Advanced Virtual Power Station (AVPS) would operate like a 'digital conductor' where the ideal orchestra will be made up of many thousands of solar PV systems, electric vehicles and household appliances such as washing machines, air conditioners, and clothes driers. Indeed, the more appliances a household voluntarily includes, the more value is added to the broader energy system and the more money

the household saves. Importantly, AVPS options will be available for households of all income levels, even including those who cannot afford solar PV, and may be provided as part of a package by energy retail businesses.

Practically, the AVPS would subtly manage appliances and electric vehicles in a way that is largely unnoticed by participating households. For example, at times of peak demand for electricity the following actions would occur automatically (although these would be preset to reflect household preferences):

- Moderation of air conditioning with negligible impact on comfort conditions
- Delaying the operation of dishwashers, washing machines and clothes driers
- Turning off pool filters and electric hot water systems
- Maximising electricity exported to the grid from solar PV and/or to on-site battery storage (if fitted)
- Reducing or turning off electric vehicle charging (or even using the car battery to channel electricity back to the grid)

To ensure that local demand matches generation capacity, the AVPS would also be capable of moderating some appliance loads when an approaching cloud bank is detected that would decrease the available solar-generation capacity.

While wind power provides low-carbon electricity, as stand-alone technology it is less reliable than traditional sources due to its dependence upon variable weather patterns.

Conversely, at lunchtime on weekdays with plenty of sunshine but low household occupancy, the AVPS would automatically bring forward the operation of thousands of appliances to make use of excess generation capacity. For instance:

- Activating dishwashers, washing machines and clothes driers
- Turning on pool filtration
- Pre-heating water
- Charging electric vehicles
- Diverting solar PV generation to on-site battery storage

It would be impossible for householders to implement the above actions manually for greatest benefit to the whole energy system. But the AVPS would provide sophisticated ways of making the task simple and convenient for customers and dependable for electricity grid managers.

AUSTRALIAN LEADERSHIP, GLOBAL NEED

Tomorrow's 'smart grid/smart home' economy presents a huge opportunity for Australia potentially worth many billions of dollars over coming decades.

Fully exploited, systems such as the AVPS that coordinate renewable generation supply with electricity demand from appliances would unlock significant energy system value. Not only would energy management be more efficient and convenient, automated mechanisms for monetising and transacting this value would also be provided for the benefit of participating households and the community at large.

These solutions will be pivotal in realising a future where electricity systems are truly low-carbon, reliable and affordable. Thanks to Australia's global leadership in this domain, the nation has an outstanding opportunity to shape the future of electricity worldwide.

MARK PATERSON is the manager of Future Electricity Partnerships at CSIRO's Energy Flagship and chairs Australia's Future Grid Forum and the Smart Grid Australia Research & Development working group.

FURTHER READING

Carvallo, A. and Cooper, J. 2011, *The Advanced Smart Grid* Artech House, Boston and London.

Fox-Penner, P. 2010, *Smart Power: Climate change, the smart grid, and the future of electric utilities*, Island Press, Washington, D.C.

NEW ENGINES, NEW FUELS, NEW ATTITUDES

To avoid catastrophic climate change, 21st century road transport must be efficient and free of fossil fuels, writes Damon Honnery

AS A VEHICLE technologist it has long amused me that the fuel consumption of the FJ Holden produced in the 1950s differs little from many similar-sized vehicles built today. If fuel consumption was a measure of technological progress, we might claim that little advancement has been made.

Newer vehicles are, of course, far safer, less polluting, higher in performance and greater in comfort than their predecessors. But apart from safety, these factors might be considered secondary to the basic purpose of a passenger vehicle, transporting passengers from one place to another. For this, the key factor is the quantity and type of energy used and like many sectors of the economy, transport now faces two equally important challenges linked by energy use: human-induced climate change and diminishing reserves of cheap oil.

Greenhouse-gas emissions from the transport sector continue to account for around a quarter of global emissions. If we are to avoid dramatic climate change, the latest climate science says we need to cut greenhouse emissions by between 50 to 85 per cent from 2000 levels by 2050. The latest inventories of global reserves and resources of fossil fuels suggest that although prices are likely to rise, we are unlikely to run out of fossil fuels within the time left to respond effectively to global warming. The prospect of the US becoming a net exporter of fossil fuels on the back of increased use of non-conventional gas sources supports this possibility.

To what extent can we expect technical innovation to overcome these increasingly urgent challenges? There are generally considered to be four options available: efficiency, fuels, power plants and demand.

EFFICIENCY

Efficiency here refers not to the efficiency of the vehicle's engine, but rather the efficiency of the vehicle itself. We normally measure vehicle energy use by how much fuel is used per unit distance driven, but since the purpose of a vehicle is to transport occupants, an alternative is to measure energy needed for each occupant over a unit of distance. Measured this way, a bus has the capacity to be more efficient than a car.

But vehicle occupancy is not the only measure. Vehicle mass is also important since energy must be used to move this as well. The lighter the vehicle the better its fuel efficiency will be and for this reason motor scooters are generally very efficient. We might, therefore, envisage future transport needs being met by buses for mass transport and small efficient motorcycle-like vehicles for personal transport.

Since many alternative fuels are derived from biomass, their production can compete with food and fibre.

Not all transport can fit into these two vehicle types, however. We will always need trucks to transport goods and large passenger cars to transport families and other small groups. Further, if fuelled by fossil fuels, even this combination might not be enough if transport demand continues to rise.

FUELS

Many countries now have mandated targets for low-carbon alternative fuels. For diesel engines, alternatives are generally based on so-called transesterified oils and fats to yield biodiesel, while for petrol engines it is alcohols such as ethanol.

For alternatives to make a difference we will have to substitute a large proportion of current fossil fuel use. Since many alternative fuels are derived from biomass, their production can compete with food and fibre. Their increased

123RF

If you measure the energy a vehicle needs for each occupant over a unit of distance, a bus has the capacity to be more efficient than a car.

use has brought about a growing realisation that large-scale production has the potential to alter commodity markets, and paradoxically, increase emissions of greenhouse gases.

Pressure on markets can come about by increased demand for raw materials and through farmers altering their crops to those used in fuel production, while increased greenhouse emissions can result from greater fertiliser use, disturbance to existing land cover and soils, and from the energy used in fuel production.

These effects can be limited if non-food sources such as woody-cellulose and grasses grown on low-grade land are used rather than sugar or corn. But such sources add considerably to fuel cost as cellulose needs to be converted to simpler compounds before conversion to alcohol. The result: we're yet to see economic production at the scale necessary for cellulosic ethanol to rival simpler sources.

That's why researchers have turned towards production of fuels more suitable for diesel vehicles. Woody-cellulose can be converted to

a diesel-like fuel via thermo-chemical conversion processes. In Australia, research is centred on use of various species of indigenous tree as the feedstock for this fuel. An added benefit is that replanting these trees in regions where they once dominated can help to improve soil quality, such as in the Western Australia wheat belt where soil salinity is a major problem.

Local and international teams are also investigating algae as a feedstock. Like many seed crops, many strains of algae contain oils that can be extracted and converted to biodiesel for diesel vehicles. Although difficult to harvest, algae promise significantly higher yields per hectare than other biomass feedstocks.

Other possible alternatives are methane and hydrogen. Methane gas, which burns easily in suitably modified engines, can be produced from a variety of sources. Only when it is produced from biomass, however, will it yield a significant reduction in greenhouse emissions.

Hydrogen's advantage is that it can be used in a fuel cell to generate electricity, or directly in

The fuel consumption of the FJ Holden produced in the 1950s differs little from many similar-sized vehicles built today.

Unless electric vehicles can be charged with power from low-emission sources, greenhouse gas emissions might be no better than with diesel or petrol engines.

a spark-ignition engine. But not only is it difficult to store in large quantities under pressure, hydrogen must be produced from a low-carbon electricity supply if it is to reduce greenhouse emissions. And unlike many alternative liquid fuels, much of the infrastructure needed to deliver gaseous transport fuels would need to be built. For hydrogen this is more complex as it can reduce the strength of the steel used in pipes.

POWER PLANTS

Beyond continuing development of existing engines a variety of new vehicle power plants are entering the market. Electronic control of fuel delivery, engine stop-start technology and variable cylinder operation exemplify how computer control has improved the fuel economy of existing engines. Additional changes come through turbocharging engines, offering greater efficiency plus lower overall engine mass for the same delivered power.

Continued improvement to both emissions and efficiency requires greater understanding of engine fuel-air mixing processes. Working with researchers from the Advanced Photon Source in Argonne National Labs in Chicago, my group uses high energy X-rays to probe high-pressure fuel sprays to determine exactly how the fuel mixes with the surrounding air. In my laboratory we use ultra-fast digital cameras, operating at up to one million frames a second, to investigate how fuel injection is altered by fuel and injector properties.

Still, existing engines will one day reach their ultimate level of efficiency. The gains we're making are gradually reducing. Overcoming these limits requires different power plants, so much effort is placed into the development of electric power in both pure and hybrid systems. Hybrid engines combine petrol and electric

motors and can convert braking energy into electricity for storage in a battery. Vehicles powered by electricity alone require large numbers of batteries to store the energy needed.

For both vehicle types much of the research effort is in developing better and better batteries. Because of their superior performance, lithium-ion has replaced nickel-cadmium as the preferred battery for electric vehicles. When used in small, lightweight vehicles, they have the potential to meet many transport needs, such as the daily commute to work.

Developments could see battery energy storage capacity double in the next 20 years. But more than greater energy storage is required to ensure widespread adoption of electric vehicles. Fast-charge infrastructure is needed to make operation easier and battery costs must be reduced. Unless the electricity used to charge them is produced from low-emission sources like wind or solar, greenhouse-gas emissions might be no better than with diesel or petrol engines.

DEMAND

Meanwhile, demand reduction emerges as a key possibility. We must travel less. Moreover, improved-efficiency fuels and power plants are unlikely to be enough. Why? Increasing demand requires rates of technological change exceeding levels that can be realistically achieved and, clearly, all technologies have limits imposed by the laws of physics and thermodynamics.

Aircraft can be made more efficient through the use of lightweight composite materials and greater streamlining.

AIR TRANSPORT

Air transport faces similar problems to road transport, except that the options are more limited. The enormous power required to lift aircraft significantly limits alternative fuel and power plants. Because of their low energy density it's not possible to use gaseous fuels or battery-powered electric motors. Engines powered by liquid fuels will be with us for a long time. Luckily, many alternative liquid fuels developed for road vehicles can be improved for use in aircraft engines. Engine efficiency can be improved using exotic materials able to withstand the increased temperatures that higher efficiency demands. The aircraft itself can be made more efficient by use of lightweight composite materials, through low drag surface treatments and greater streamlining. But, just as with land transport, all these improvements must be seen in the context of increasing demand for national and international air travel.

Dramatically reducing transport needs is no easy task. It might ultimately require us to redesign our cities and urban areas to make them more amenable to walking and using non-motorised vehicles such as bicycles.

Increased bicycle use in many Australian cities and the resulting need for greater traffic management to make this easier and safer is one example of how this is occurring.

Whatever the future holds for road vehicles, meeting the challenges of climate change and diminishing reserves of cheap oil will require us to break free of the constraints imposed by our existing vehicles. The research being done on alternative fuels, electric vehicles and on demand management suggests this might be possible; we might finally break free of the FJ Holden.

ASSOCIATE PROFESSOR DAMON HONNERY is an associate professor in the Department of Mechanical and Aerospace Engineering at Monash University.

FURTHER READING

Moriarty, P. and Honnery, D. 2011, *Rise and Fall of the Carbon Civilisation*, Springer, London.

Moriarty, P and Honnery, D. 2013, 'The global environmental crisis of transport', in Low, N. (ed) *Transforming Urban Transport, the Ethics, Politics and Practices of Sustainable Mobility*, Earthscan Routledge, Oxford and New York.

Kahn Ribeiro, S. et al 2007: 'Transport and its infrastructure', in *Climate Change 2007: Mitigation*, in Fourth Assessment Report of the Intergovernmental Panel on Climate Change, Metz, B. et al (eds), Cambridge University Press, Cambridge and New York.

James, M. 2009, 'The (green) car of the future', Australian Parliamentary Library Background Note, http://www.aph.gov.au/About_Parliament/Parliamentary_Departments/Parliamentary_Library/pubs/BN/0910/CarOfTheFuture.

IS ANYBODY OUT THERE?

ET hasn't called home yet, but scientists are searching for cosmic company and Earth-like planets, writes Jude Dineley

NASA

NASA

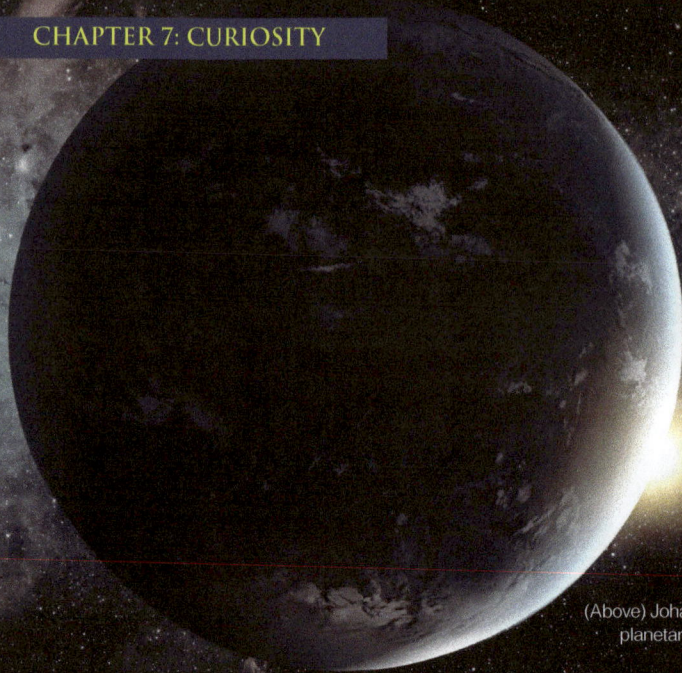

(Above) Johannes Kepler, the 17th century astronomer whose laws of planetary motion are part of the foundation of modern astronomy. (Left) An artist's impression of the Kepler-16 system, which has two suns, pictured behind planet Kepler-16b.

THE HAIRS WERE standing up on the back of my neck," says American astronomer Geoff Marcy of the moment he realised he had discovered one of the first planets outside our Solar System.

Marcy and his postdoctoral fellow Paul Butler were analysing data at the University of California at Berkeley during the Christmas holiday of 1995 when evidence of 70 Virginis B jumped off the computer screen. They had discovered a giant gas planet more than six times the mass of Jupiter orbiting a Sun-like star 72 light-years away. Months earlier, researchers in Geneva had found the first planet orbiting a Sun-like star outside of the Solar System, 51 Pegasi B, 50 light-years – 9.5 trillion kilometres – from Earth.

It was a giant leap forward in the hunt for planets beyond our Solar System – known as exoplanets – and a small step forward in the search for extra-terrestrial life.

Since then, many teams have joined the search, confirming the existence of more than 890 exoplanets, a tally that climbs on an almost weekly basis. It's an incredible transformation of the field that 20 years ago was viewed by many astronomers as a kooky quest.

The discoveries have overturned conventional thinking of what other planetary systems look like. Until the 1990s, astronomers assumed that all systems were like our Solar System. In fact, a diverse range of planet sizes and orbits has been revealed, some of them orbiting more than one star.

"It turns out that almost no other Solar System is like ours," says Chris Tinney, an exoplanet specialist at the University of New South Wales.

HUNTING PLANETS

Planets like our own are of particular interest to astronomers – since the only life we know is found here on Earth, rocky planets with similar environments are thought to be our best bet in the search for extra-terrestrial life.

But finding rocky Earth-like planets is tricky. It's near impossible to spot these planets directly with telescopes, as the stars they orbit are dramatically bigger and brighter than them. Instead, they are detected indirectly by measuring the effects they have on their host star. Two approaches are responsible for more than 90 per cent of the discoveries to date: one technique detects the gravitational 'wobble' of

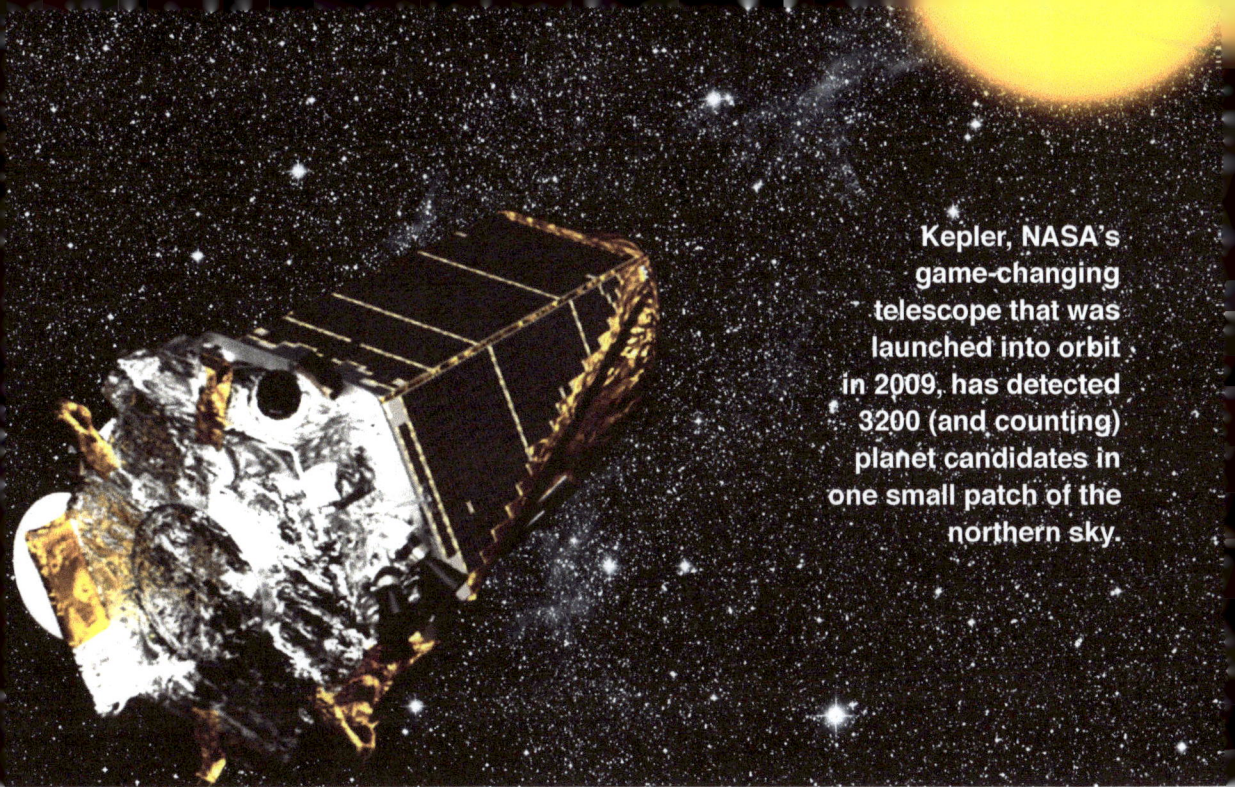

Kepler, NASA's game-changing telescope that was launched into orbit in 2009, has detected 3200 (and counting) planet candidates in one small patch of the northern sky.

a star produced as a planet orbits around it, pulling it back and forth; the other, so-called transit approach measures the 'dimming' planets cause in their parent star as they pass in front of them, orbit after orbit.

Among several sites using the 'wobble' technique worldwide, Australia has its own planet hunter at Siding Spring Observatory in northern New South Wales. In a search led by Tinney since 2000, the 3.9-metre Anglo-Australian Telescope (AAT) has discovered more than 40 planets.

The AAT is also following up transit measurements made by a network of southern hemisphere telescopes called HAT-South that include one at Siding Spring. While transits indicate the size of a planet, 'wobble' measurements provide an estimate of mass.

"The combination of those two [techniques] is incredibly powerful because it gives you the density of the planet and that then tells you whether it is a rocky planet, like Earth, or gas giant planet, like Jupiter," says Tinney. He and his group have found mostly gas giants this way, but as telescope precision improves, they hope to find smaller and smaller planets.

Nevertheless, for telescopes on the ground

using the transit approach the odds are against finding Earth-sized planets, which dim a star by only 0.01 of a per cent. The Earth's atmosphere reduces and distorts the light reaching the telescopes, making detection even trickier.

Enter Kepler, NASA's game-changing telescope that was launched into orbit in 2009, where the light it detects is not distorted by the Earth's atmosphere. Trailing the Earth around the Sun, the telescope has detected 3200 (and counting) planet candidates in one small patch of the northern sky. Marcy, a co-investigator on the mission, says the tally includes hundreds of Earth-sized planets.

"Kepler has completely changed our perspective about how common Earth-sized planets are," he says. "We had no idea of this three years ago."

While mechanical failures recently ended Kepler's original mission, two new space-borne telescopes, including NASA's Transiting Exoplanet Survey Satellite (TESS), are scheduled for launch in 2017.

Unlike Kepler, TESS will scan the entire sky, investigating the nearest and brightest stars. This is exciting news for astronomers in Australia, enabling them to probe the

The Anglo-Australian Telescope at the Siding Spring Observatory in New South Wales has been undertaking an astronomical survey since 1998 and has found more than 40 planets.

southern sky in more detail. "There's a whole range of follow-up observations that become much more doable when the stars are 1000 times brighter than the stars Kepler is looking at," explains Tinney.

PLACES LIKE HOME

Finding rocky Earth-sized planets is no guarantee that life will also be found. After all, close neighbour Venus is also a terrestrial planet and close in size to Earth, but is positively hostile to life. High carbon dioxide levels result in an extreme greenhouse effect that generates scorching temperatures of 400°C at its surface. The high temperatures mean that the surface of Venus is missing one important ingredient – liquid water.

The only life we know, that here on Earth, is dependent on liquid water and conventional thinking is that extra-terrestrial life needs liquid water too. For water to be present, a minimum requirement is that a planet must sit in a 'habitable' or so-called 'Goldilocks' zone, not so close to its star that the star's heat boils any water away, but not so far away that it freezes.

The habitable-zone concept isn't bulletproof by any means. Recent research has shown that Venus does in fact sit in the Sun's habitable zone, yet it has no liquid water. Neither does Mars, which also sits at the 'just right' distance from the Sun. Its mass is only 11 per cent of that of Earth and the resulting lack of

THE MARTIAN OUTBACK

Seeing the Pilbara's scorched red landscape in Western Australia, you could be forgiven for thinking it looks otherworldly. The region has close parallels with Mars. Three-and-a-half billion years ago, its climate and landscape had much in common with that of our Solar System neighbour.

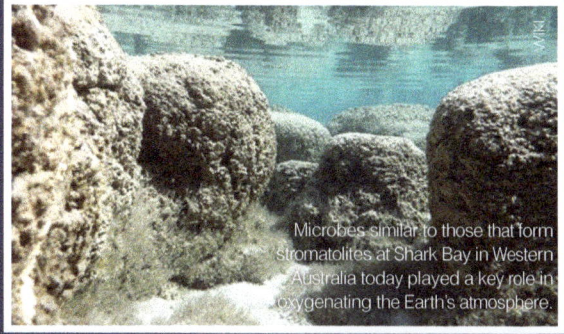

Microbes similar to those that form stromatolites at Shark Bay in Western Australia today played a key role in oxygenating the Earth's atmosphere.

At that time, both locations had hot, wet climates and landscapes that included oceans, lakes and rivers, an environment that on Earth at least supported diverse communities of microbial life. Fossilised remains of that life found in the region are some of the most ancient known to man.

"We can make a direct comparison between Earth and Mars and predict whether there might have been life on Mars at the same time," says Malcolm Walter, a geologist and astrobiologist at the University of New South Wales.

The microbes on Earth formed multilayered rock sandwiches called stromatolites, where layers of the microbes are interleaved with sheets of sediment, trapped by mucus they secrete. Using photosynthesis, the microbes produced the first oxygen in the Earth's atmosphere, allowing more complex organisms to flourish.

Today, fossilised remains are found at sites across the Pilbara and living microbial populations thrive in Hamelin Pool in Shark Bay – one of two locations that host the most diverse living stromatolite populations on Earth.

These living and fossilised microbial communities provide the ideal test bed for developing the science used to detect similar life on Mars. In fact, strategies used on board NASA explorers Spirit and Opportunity were tested in the Pilbara.

Research has included important development work that enables the stromatolites and microfossils to be correctly identified, as they can also resemble rock features formed by chemical reactions.

Walter sees Mars as our best chance for finding extra-terrestrial life, given its close proximity and favourable environment: "There is a very good chance that there is, or at least was, microbial life on Mars at an early stage."

gravitational pull means it has lost any protective atmosphere that it once had, plunging its surface to temperatures at which liquid water cannot exist.

Nevertheless, the habitable zone is still a guiding principle that astronomers can use to narrow down the hunt for potentially life-friendly planets.

The smallest planets yet found in a habitable zone were confirmed this year orbiting the star Kepler-62. The super-Earths – planets between two and 10 times the mass of Earth – Kepler-62e and Kepler-62f are 60 per cent and 40 per cent bigger than the Earth and initial data suggests that Kepler-62f is a rocky planet, like Earth.

In the first estimates of their kind, Kepler researchers, including Marcy, have proposed that the Milky Way has tens of billions of potentially life-friendly Earth-like planets and that some of them are likely to be close by, in galactic terms.

"Imagine the whole galaxy shrunk down to the size of Australia," says Marcy. "The distance from us to the nearest habitable planet would be the distance from central Sydney to the Opera House."

FINDING LIFE

With the proposed tens of billions of habitable zone planets in our galaxy alone, it would seem the chances for life in our galaxy are good.

"If you buy 10 billion lottery tickets not all of them are going to be losers," says Seth Shostak, an astronomer at the Search for Extraterrestrial Intelligence (SETI)

An artist's impression of Kepler-62f and its neighbour Kepler-62e, the planets closest in size to Earth found in a habitable zone.

Institute in Mountain View, California.

Considering the unimaginably large scale of the universe, where our Sun is one of around 200 billion stars in our own Milky Way, and the Milky Way is one of an estimated several hundred billion galaxies, the chances of life in the wider universe are higher still.

Of three main strategies for finding life on other planets, a search of Mars arguably offers the best chance of success (see: **The Martian outback)**.

A second line of investigation is a search for the chemical signatures of life in the atmospheres of distant planets. The detection of oxygen, for example, could indicate the presence of photosynthesising life forms. But according to Tinney, it could be decades before astronomers have technology that can convincingly do the job. On top of that, many biosignature chemicals can be produced without life, raising the risk of false positive discoveries.

A third strategy, and something of a wildcard in the hunt for life, is to use existing telescopes to look for more obvious signs associated with advanced civilisations.

Signs fall into two main categories: communication signals and alien-engineered structures that could include giant space-borne solar panel networks called Dyson Spheres, imagined to be built to meet the high energy demands of an advanced life form.

In new work for SETI, Marcy is leading projects that will analyse Kepler transit data for unusual dimming patterns that cannot be explained by natural phenomena and search for laser transmissions from other civilisations.

According to Shostak, Kepler-62e and Kepler-62f, together, are a golden opportunity for a targeted search for transmissions. If life exists on both, they could be talking back and forth.

"If you look in that direction when the two planets are lined up with Earth you're looking right down the communication pipeline between those two planets," he says.

Marcy is pragmatic about whether aliens will be found on his watch but notes: "If you don't look, you certainly won't find them." Recalling his hair-raising discoveries of 1995, he adds, "I've been lucky once, maybe I'll be lucky again."

DR JUDE DINELEY has a doctorate in medical physics and is a freelance science writer based in Sydney.

FURTHER READING

NASA Exoplanet Archive, website, http://exoplanetarchive.ipac.caltech.edu/.

Kepler mission homepage, website, http://kepler.nasa.gov/.

A virtual field trip to see the Shark Bay stromatolites, http://vft.asu.edu/.

The Great Exoplanet Debate, *Astrobiology* magazine, hosted during the 2012 Astrobiology Science Conference in Atlanta, Georgia, http://www.astrobio.net/debates/10/the-great-exoplanet-debate.

COSMIC QUEST

Scientists are revealing the origins and structure of the universe, black holes, dark energy and dark matter, writes Wilson da Silva.

I'S THE ULTIMATE question: where did we come from? How did we get here? These are overwhelming questions that have probably been asked as long as there have been people. Today, scientists working in this field, known as cosmology – the study of the origin, structure and dynamics of the universe and its processes – are finally answering them. And Australians are at the forefront of both finding the answers and developing the astonishingly complex technology required to find them.

Only in the 20th century has cosmology improved on what the Ancient Greeks knew. Since Galileo's time 400 years ago, people have known that some 'stars' – those that wandered the night sky – were actually planets, that they were much closer to us and, like our Earth, they orbited the Sun.

A century later, Isaac Newton developed a powerful explanation for their movements based on how falling objects behave. These laws of motion and universal gravitation apply not only to falling apples, but to how planets move in the sky. The laws came to dominate our understanding of the physical universe for the next 300 years.

When the 20th century began, the Milky Way galaxy – the bright swathe of stars easily seen at night – was thought to be our entire universe. But as more powerful telescopes were built, astronomers were astounded to discover that the universe was more immense than anyone had imagined.

It took until the late 1920s to accept fully that the Milky Way was just one of many 'island universes', or galaxies. For thousands of years, scientists had believed the universe was static and unchanging. But suddenly, there was a lot more universe than anyone expected.

And then it got really strange.

THE SPEED OF LIGHT

In a scientific paper published in 1905, Albert Einstein proposed his theory of special relativity, the now famous $E=mc^2$. In it, he expanded Newton's laws of motion so that they applied to objects moving at high speed and explained why light was unaffected by these laws: why it was that, if a car travels at 80 km/h, its headlight beams do not travel at the speed of light plus 80 km/h.

The Voyager program's spacecrafts, launched in 1977, gave us an unprecedented look at Jupiter, portrayed here with four of its satellites – Io, Europa, Ganymede and Callisto.

He postulated that the speed of light is the same for all observers, and set the speed of light in a vacuum – 299 792 458 metres per second – as a universal constant; the maximum speed at which all energy, matter and information in the universe can travel.

Later, in 1916, Einstein postulated the general theory of relativity, applying his earlier idea to larger bodies governed by Newton's laws of universal gravitation. In Newton's model, gravity is the result of an 'unknown force' produced by immense objects acting on the mass of other objects.

Einstein suggested gravity arose as a property of space and time: the greater the mass of an object – such as a moon or a sun – the more it distorted, or bent, the fabric of space. And because relativity links mass with energy, and energy with momentum, the curvature of space-time – the gravitational effect – was directly related to the energy and momentum of whatever matter and radiation was present.

It sounded bizarre, but neatly accounted for several strange effects unexplained by Newton's law, such as anomalies in the orbits of Mercury and other planets. If true, it would also mean the universe is not static, something Einstein himself worried about. So much so, that he added a positive 'cosmological constant' to his equations to counteract the attractive effects of gravity on ordinary matter, which would otherwise cause the universe to either collapse or expand forever.

But he didn't need to worry. American astronomer Edwin Hubble astounded the world in a 1929 paper showing conclusively that the further away astronomical objects are, the faster they appear to be travelling away from us. The only possible explanation was that the universe was expanding and, hence, changing – just as Einstein predicted.

In fact, Hubble's data meant the universe itself had a beginning and an end – a seemingly fantastical prediction made by Belgian astronomer Georges Lemaître only two years earlier. This was the birth of what we know today as the Big Bang theory.

After the initial expansion, about 13.7 billion years ago, the universe cooled sufficiently to allow the superheated energy to condense into various subatomic particles, eventually forming protons, neutrons and electrons. From this morass came the simple elements of hydrogen, helium and lithium. Gigantic clouds of these elements coalesced as gravity took hold, forming stars which ignited with heat and light, eventually forming galaxies as the stars clumped toward each other.

Inside stars, clouds of gas – compressed by the titanic crush of gas columns above them – created all the heavier elements: carbon, iron, silicon, lead and so on. Some stars eventually died in cataclysmic explosions known as supernovae, expelling heavier elements into space where gravity again brought them together, forming planets and moons. Even the "echo of the Big Bang" was discovered in 1965 – a cosmic background radiation permeating all space, created by the raging oceans of white-hot energy at the dawn of time.

A SHIFT IN THINKING

From the 1960s on, astronomers tried to answer the question: would the universe expand forever or collapse in a Big Crunch?

They didn't get far. Observations suggested there wasn't enough visible matter in the universe to account for gravitational forces they observed acting within and between galaxies. Among the

WIFI FROM SPACE

Another surprising repercussion of Einstein's theories was that a stellar object might grow so massive that even the speed of light was not fast enough to escape its gravitational pull – theoretical objects dubbed 'black holes'. But they would be devilishly difficult to find. In 1974, physicist Stephen Hawking suggested that under certain circumstances, small black holes might 'evaporate' and leak radio signals as they vanished. These signals would be weak, buried in background cosmic noise and probably 'smeared'. In 1983, hoping to be the first to detect an evaporating black hole, CSIRO physicist and engineer John O'Sullivan and his colleagues came up with a mathematical tool to detect the tiny, smeared signals against a background of intergalactic distortion.

They didn't find them. But the technique they developed, he realised, could allow data sent wirelessly over many different frequencies to be recombined at the receiver. And so, WiFi was born.

first to recognise this was Australian astrophysicist Ken Freeman, whose 1970 paper on how spiral galaxies rotate started the shift that came to be known as 'dark matter':

"There must be in these galaxies additional matter which is undetected … its mass must be at least as large as the mass of the detected galaxies, and its distribution must be quite different from the exponential distribution which holds for the optical galaxy."

The paper is now one of the most cited single-author papers in astrophysics.

Suddenly, a big portion of the stuff making up the universe was understood as a strange kind of matter that doesn't emit light or interact with normal matter in any way except via gravity. Evidence for dark matter has since been observed countless times: in the lumpiness of large sections of space – the way thousands of galaxies tend to

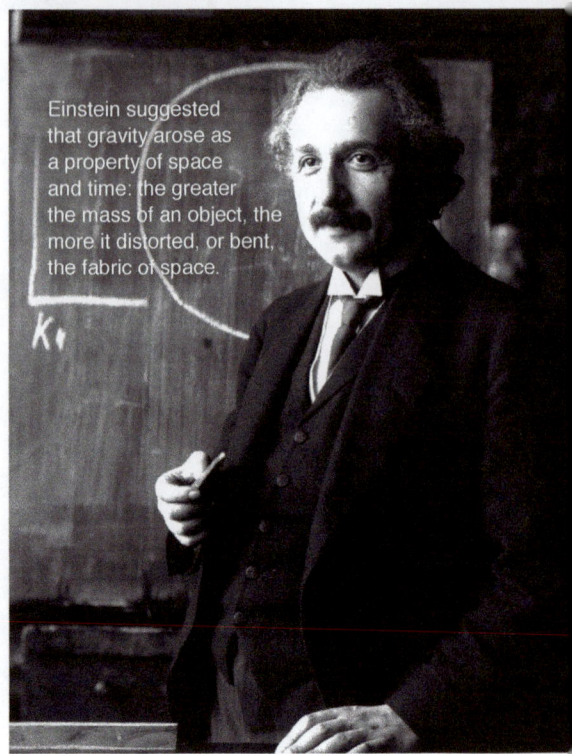

Einstein suggested that gravity arose as a property of space and time: the greater the mass of an object, the more it distorted, or bent, the fabric of space.

Understanding just 5 per cent of the physical universe has already brought astonishing advances in technology.

clump together, which suggests something invisible is drawing them closer; or the velocity dispersion of galaxy clusters (how fast stars move relative to each other), which give clues about their mass – and that mass is greater than the mass of visible entities.

DARK MATTER AND DARK ENERGY

So what is dark matter? Physicists have proposed a number of new particles to account for it. One group is known collectively as weakly interacting massive particles (or WIMPs), another is a new light neutral particle, the axion. Several projects to detect them directly are underway – mostly in deep underground laboratories that reduce the background effect from cosmic rays – in places like old iron and nickel mines in the US and Canada and in a mountainside in Italy.

As if that wasn't enough, astronomers were stunned in 1998 when two international teams – one led by Australian Nobel Prize-winning astrophysicist Brian Schmidt – announced that the expansion of the universe was *accelerating*, rather than slowing as had been expected. This was a surprise as gravity slows moving objects over time: hence, the Universe's expansion should be slowing in the billions of years since the Big Bang.

Einstein's general theory of relativity allows gravity to push as well as pull, but most physicists had thought this purely theoretical. Not any more.

How gravity does this and what does it, is a complete mystery. One explanation is that 'dark energy' is a property of space that possesses energy. Because this energy is a property of space itself, it wouldn't be diluted as space expands. As more space is created by the expansion of the universe, more dark energy is created, causing the universe to expand faster and faster.

What cosmologists now know is that 68 per cent of the universe consists of dark energy, and 27 per cent is dark matter. The rest is 'normal' matter: everything we can see and touch, and everything visible beyond Earth – and it makes up less than 5 per cent.

It's amazing to think how far cosmology has come in a century and how much more is known today than in 99 per cent of human history. Understanding just 5 per cent of the physical universe over the past four centuries has already brought astonishing advances in technology and living standards. Whatever the explanation for the puzzling 95 per cent that remains, it will surely lead to important new insights equally as beneficial.

WILSON da SILVA is a science journalist in Sydney, and the co-founder and former editor-in-chief of the science magazine *Cosmos*.

FURTHER READING

Weinberg, S. 2008, *Cosmology*, Oxford University Press, Oxford.

Krauss, L.M. 2012, *A Universe from Nothing: Why there is something rather than nothing*, Free Press, New York.

Sagan, C. 1980, *Cosmos*, Random House, New York.

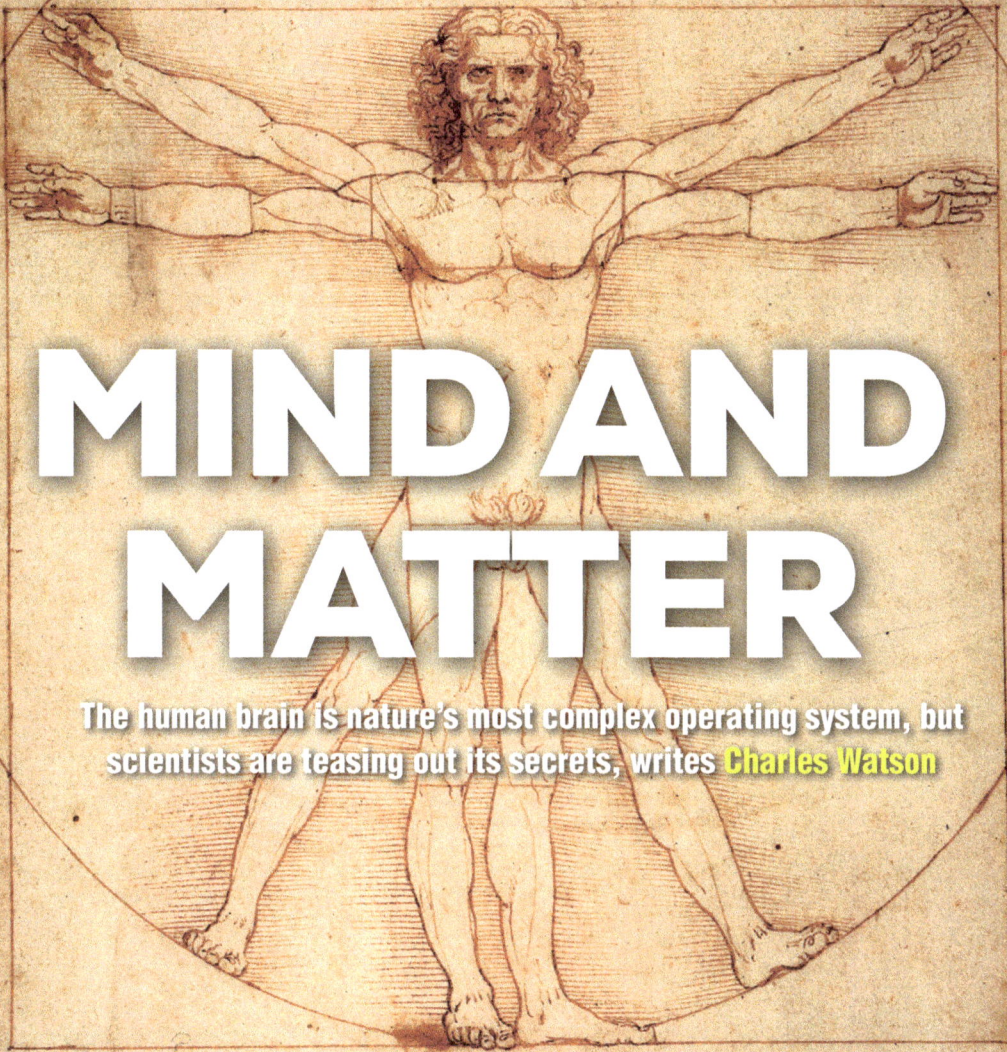

MIND AND MATTER

The human brain is nature's most complex operating system, but scientists are teasing out its secrets, writes Charles Watson

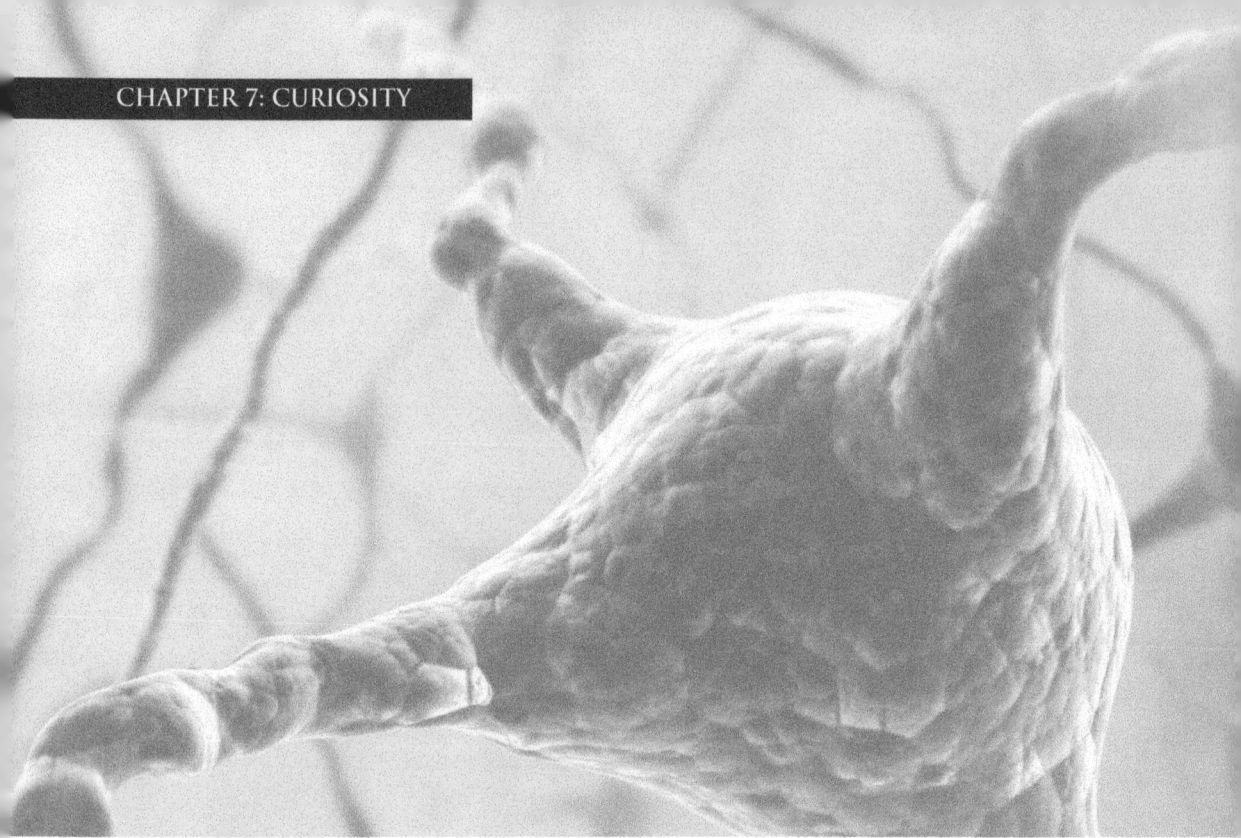

WHAT DO HUMAN brains and insect brains have in common? Quite a lot! A busy cockroach can find food, escape from danger, and produce offspring with a brain the size of a pinhead.

Insect brains have structures surprisingly similar to the basic motor control systems of the mammalian forebrain and the nerve cord of insects develops under the control of the same genes that make our brain stem and spinal cord. It looks as if the basic components of a brain were developed in very early living animals, and that the major elements have been retained ever since then. Researchers at the Queensland Brain Institute have learned a great deal from the brains of bees and fruit flies.

This isn't really surprising. Insect and mammal brains like ours have the same fundamental purpose, survival: of the individual and the species. Brains code information from the outside world and from internal sensors into millions of signals that can be processed and interpreted by networks of cells called neurons.

The aim is to choose the best possible responses by the body's muscles and glands when challenged. The brain ensures that the individual survives by getting it to eat and drink when it senses hunger or thirst, and to defend itself from attack by other animals when it perceives an approaching threat. In vertebrates, the brain ensures that the species survives by encouraging sexual reproduction and care of the offspring. In humans, these survival functions are coordinated by the hypothalamus, a very small region only about 1 cubic centimetre in volume.

On the input side, the hypothalamus is informed by messages from all of the senses and memories of important past events. This part of the brain can initiate movements and control hormone systems. The movement control systems are organised very economically. Important movement sequences such as running, grasping, licking, and chewing are encoded in parts of the brain in much the same way as computers use subroutines, enabling each movement to be ordered with a single command.

Depending on the need, different combinations of movement modules are enlisted. For example, hunger in a rat will trigger looking for food,

Brains code information from the outside world and from internal sensors into millions of signals that can be processed and interpreted by networks of cells called neurons.

Human and insect brains have quite a lot in common. The nerve cord of insects develops under the control of the same genes that make our brain stem and spinal cord.

running to find it, grasping, licking, chewing and swallowing it. Using its control of endocrine glands and internal organs, the hypothalamus is able to maintain a stable internal environment (such as a constant temperature) in the face of external challenges. This function is called homeostasis. The Herzog group at the Garvan Institute are unravelling the homeostatic control of appetite and obesity.

The human brain has the same survival systems as other mammals and a great deal of what we do every day uses these simple systems. The difference between simple brains and the human brain is that we have a huge cerebral cortex sitting on top of the hypothalamus, giving us a much wider range of choices about the way we live our daily lives.

Researchers have learned much about the human brain by looking at our recent evolutionary history. Humans evolved from the same distant ancestor as monkeys and apes and with them we form a group known as primates. Among the special features of the primate brain are elaborate systems for accurate vision and precise control of hand movement. We inherited these features and have developed a very

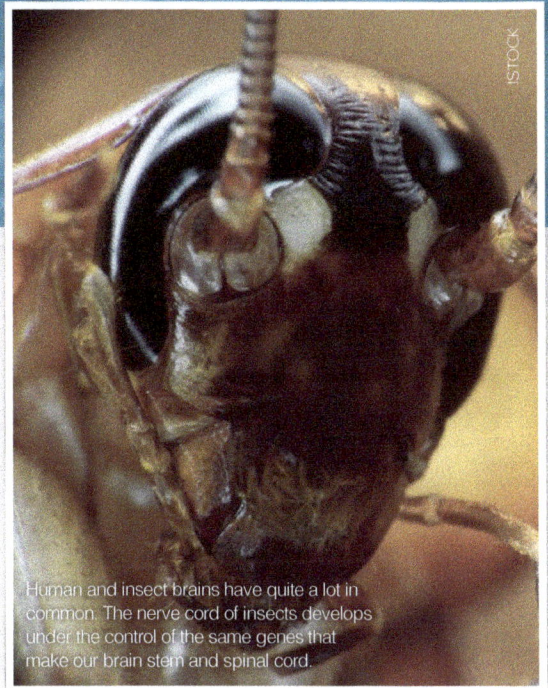

powerful capacity to take advantage of them – the power of complex communication.

All animals communicate at a basic level, mostly through vocalisation and body language. But over the past three million years the human brain has developed a system of communication with almost infinite potential that uses not only vocalisation and body language, but also written and electronic forms of communication. This has led to an explosion in knowledge transfer.

Three million years ago early humans walked upright, lived and hunted in groups, and used

MACHINE BRAINS

The film *2001: A Space Odyssey*, with the very human-like fictional computer named HAL, kick-started a fascination with the prospect of Artificial Intelligence, with computers so complex they could operate like a human brain. While computer simulation of human brains is in question, there's no doubt that computers are enormously powerful in their own way, and can outperform humans in very complex tasks. IBM's 'Watson' computer proved that it could beat humans in quiz contests, and is now being used to assist cancer diagnosis and treatment.

The IBM-sponsored Blue Brain project in Switzerland has attempted to build a computer model of a basic element of rat cerebral cortex, a small cylinder of cortex consisting of 10 000 neurons. The project's successor, the one-billion euro Human Brain Project, aims to create thousands of computer replicas of this cortical cylinder and link them together. The project was announced in January 2013 and will be directed by the Swiss Federal Institutes of Technology in Zurich. It seeks to reveal how the human brain works by building a silicon version in a supercomputer.

The BRAIN Initiative, launched in April 2013 by President Barack Obama, will not focus primarily on computer modelling of the brain, but will attempt to develop technologies for mapping the activity of tens of thousands of neurons simultaneously. This is still very ambitious.

Despite advances in simulating human thought processes, the operating system of the human brain is fundamentally different from the way computers work. Until we understand that operating system of the human brain, computer simulation may be shooting in the dark.

primitive stone tools. The development of communication through speech was highly advantageous because it assisted the group in hunting and defence and ensured that any invention was shared between all members.

Complex communication led to the formation of groups with strong social bonds and an ability to share duties of food gathering, defence, and child rearing. In evolutionary terms, the advantages of sophisticated communication skills were so great that every step in brain expansion was exploited to the benefit of each human society. The result was continuous expansion in the number of neurons in the cerebral cortex, from 30 billion in our earliest ancestors to around 86 billion in modern humans.

Counting neurons isn't easy. The number of neurons in the human brain has been the subject of debate for decades. Fortunately, Brazilian neuroscientist Associate Professor Suzana Herculano-Houzel, with the Federal University in Rio de Janeiro, developed a simple way of accurately counting neurons in 2005. She puts a brain in a blender with detergent to release the nuclei from all the brain's cells. The intact nuclei of neuronal cells and non-neuronal glial cells (glial cells are the supporting cells of the nervous system – they help to keep neurons alive) can be distinguished with special staining, and the counting can be done quickly and accurately with a standard red blood cell counting system.

There's an interesting culinary twist in the story of human brain expansion. While early humans were skilled food gatherers, using stone tools to strip carcasses and break bones to get marrow, their brains would probably have stayed at about the 40 billion neuron mark if it weren't for the invention of fire. Fire was crucially important because cooked food is about four to five times more nutritious than raw food.

Gorillas and chimpanzees, with a brain the same size as our early ancestors, need to spend six to eight hours a day gathering food. The brain is a very greedy organ. Although it is only 2 per cent of body weight, it uses more than 20 per cent of the energy from food. Without fire and cooking, brain evolution would have stalled because there weren't enough hours in the day to collect more food to feed a bigger brain.

Apart from its sheer size, the human cerebral cortex has many more specialised areas than most other mammals. Rats, for instance, have about 20 centres that receive sensory information, give commands for movement and store memories, but which don't do much else. In contrast, the human cerebral cortex evolved roughly 200 distinct areas, most of which are engaged in analysis, understanding and communication, rather than simple sensory reception and control of movement.

Go back to that cockroach brain. There's no doubt our elaborate brain beats the insect brain hands, or neurons, down. Scientists have used their own big brains to learn much about brain structure and function. Still, there's a long way to go. The main limitation is a poor understanding of the operating principles of the *Homo sapiens* cerebral cortex.

Meanwhile, Australian neuroscientists at the Florey Institute, University of Melbourne, and the Brain Mind Institute at the University of Sydney are focusing on identifying the underlying mechanisms of degenerative brain conditions such as Parkinson's and Alzheimer's diseases. The goal is early treatment of these debilitating disorders.

This isn't the same as understanding how the whole brain works, but progress here promises enormous impact worldwide. At a basic science level, the researchers at Neuroscience Research Australia at the University of New South Wales

Homo habilis was the earliest of our ancestors to show a significant increase in brain size.

are the world leaders in brain mapping, which is essential for clinical studies.

Where to next? Intense research in Australia on early treatments for Alzheimer's, Parkinson's, and multiple sclerosis will deliver results over the next five to 10 years. The prospect of gene therapy for brain conditions is now just starting to open up.

PROFESSOR CHARLES WATSON is a neuroanatomist at Curtin University. He is a co-author of a brain atlas that is the most cited publication in all of brain research.

FURTHER READING

Herculano-Houzel, S., Lent, R. 2005, 'Isotropic fractionator: A simple, rapid method for the quantification of total cell and neuron numbers in the brain', *Journal of Neuroscience* 25(10): 2518-2521.

Herculano-Houzel, S. 2012, 'The remarkable, yet not extraordinary, human brain as a scaled-up primate brain and its associated cost', *Proceedings of the National Academy of Sciences of the United States of America* 109(Supp 1); 10661-10668.

Strausfeld, N.J., Hirth, F. 2013, 'Deep homology of arthropod central complex and vertebrate basal ganglia', *Science* 340(6129): 157-161.

Watson, C., Kirkcaldie, M., Paxinos, G. 2010, *The Brain: An introduction to functional neuroanatomy*, Elsevier Academic Press, San Diego.

Wrangham, R. 2010, *Catching Fire: How cooking made us human*, Basic Books, New York.

STEM CELLS: HYPE OR HOPE

Australian scientists are on the front line of the global quest for treatments based on the body's most remarkable cells, writes Elizabeth Finkel

IN 1998 SCIENTISTS in America and Australia captured the world's imagination when they created the first stem cells from human embryos.

Like the five-day-old human embryos they came from, these cells, called human embryonic stem (ES) cells, have the potential to make human flesh and blood. For millions of people with degenerative illnesses they offered the promise of spare parts and in unlimited quantities – embryonic stem cells can multiply indefinitely.

But standing in the path of that promise lay major obstacles. For starters, ethical concerns from some groups have led to opposition to the use of these cells because obtaining them requires the destruction of human embryos, even though these embryos were invariably surplus embryos that were destined to be discarded. In the US this led to tight restrictions on the use of federal funding for the research, while Italy, France and Germany banned the derivation of embryonic stem cells from embryos.

Researchers have now successfully cultured stem cells from both embryos and adult sources. Embryonic stem cells can generate any tissue. Adult stem cells form a limited number of cell types.

Australia found a balance that allowed researchers to derive embryonic stem cells from surplus embryos if they had obtained informed consent from the embryo donors and the research project had been approved both by an institutional ethics committee and a government licensing committee.

But there were issues beyond the policy obstacles. The cells were extremely difficult to control. It was hard to direct them to produce a particular type of tissue and they had an alarming tendency to produce cancers when grafted into lab mice.

Despite these problems, 15 years since their discovery, embryonic stem cells are realising their promise. Even a reserved and measured scientist such as Professor Martin Pera, program leader of Stem Cells Australia, has commented that "the future is now".

The main reason for Pera's claim is that therapies derived from embryonic stem cells are at last in clinical trials. The first trial was for spinal cord injury by US company Geron. In 2010, four patients had their spinal cords injected with cells called oligodendrocytes, which are capable of re-insulating damaged spinal cords. Within a year the trial was halted, but not because of safety concerns. Rather, the prospect of a long and costly clinical trial had dampened the investors' enthusiasm and they placed their bets on developing a cancer therapy instead.

The next two stem cell therapy candidates to run the gauntlet of clinical trials focused on blindness and diabetes.

The first has taken aim at age-related macular degeneration, the leading cause of blindness in affluent countries. Twelve per cent of Australians over the age of 50 are affected, and for the so-called dry form there is no current treatment. The primary cause of the disease is the degeneration of a specific layer of nurturing cells that lie just outside the retina of the eye called retinal pigmented epithelial cells. The good news is embryonic stem cells readily make this cell type. Replacing them is a relatively simple surgical procedure and foreign cells are tolerated by the eye.

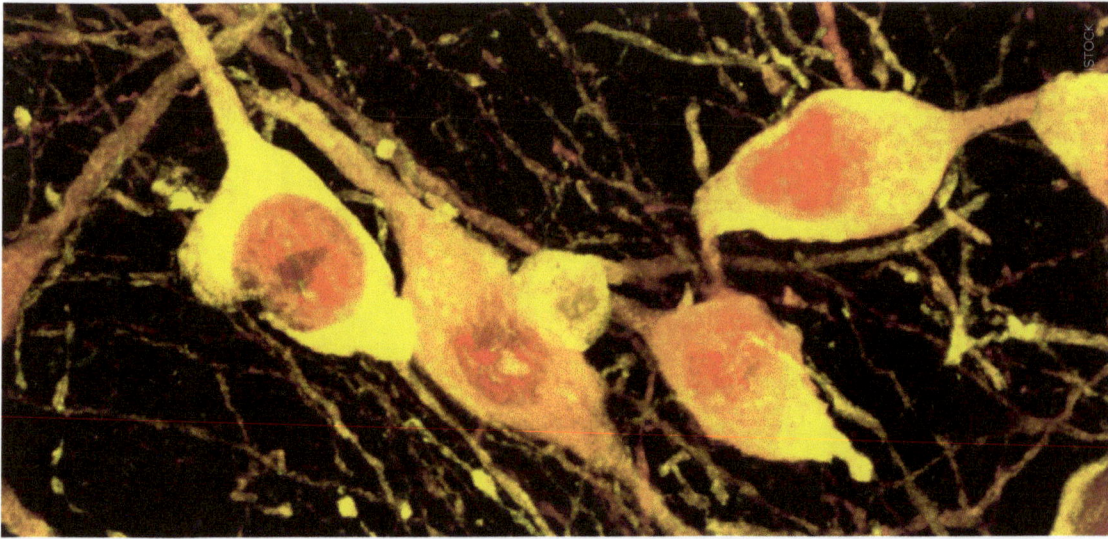

By manipulating human embryonic stem cells, Dr Lachlan Thompson at Melbourne's Florey Institute of Neuroscience and Mental Health generated neurons that produce dopamine. He grafted these into the brain of a rat suffering from the equivalent of Parkinson's disease and found the symptoms were partly alleviated. Experiments like this offer new hope to human sufferers of the disease.

TYPES OF STEM CELLS

Embryonic stem cells: derived from five-day-old embryos, divide indefinitely, can generate any tissue. They are said to be pluripotent.

Adult stem cells: present in most mature organs, have a limited ability to divide; can form a limited number of cell types. They are said to be multipotent.

Embryonic stem cells derived via therapeutic cloning (also known as somatic cell nuclear transfer): the nucleus (which carries the DNA) of a skin cell is slipped into an egg whose own nucleus has been removed. The egg now develops into an embryo from which embryonic stem cells are harvested. When it comes to potential tissue grafts, these stem cells offer a perfect genetic match to the skin cell donor.

Induced pluripotent stem cells (iPS cells): starting from a skin cell or some other mature cell, four genes are introduced. These reprogram the mature cell to a state very similar to but not identical to an embryonic stem cell. iPS cells can make any tissue type and are therefore described as pluripotent.

Several groups are developing ES cell treatments for macular degeneration, with US company Advanced Cell Technologies the front runner. One early stage trial confirmed the procedure is safe, and one patient's vision improved dramatically from 20/400 to 20/40.

Stem cell research also offers hope for people with type 1 diabetes, a disease which strikes somewhere between childhood and early adulthood. It destroys the pancreatic beta cells which produce insulin, a hormone that manages the body's sugar levels. Diabetic patients must inject themselves with a dose of insulin but it's a tough balancing act. Too much insulin and blood sugar levels plummet, potentially triggering a blackout; too little insulin and blood sugar levels rise too high, poisoning blood vessels and leading to blindness and kidney failure.

Grafts of pancreatic beta cells derived from embryonic stem cells may do a better job. Following success in animal trials, US-based firm ViaCyte Inc. has developed a credit card-sized graft for insertion beneath the skin of a patient's back. A membrane protects the graft from being attacked by immune cells but allows nutrients to pass through, providing a source of insulin to the body as it is required. Trials of this treatment in human patients are expected to start in the US in early 2014.

THERAPEUTIC CLONING

In May 2013, the world was shaken by the announcement that researchers at Oregon Health & Science University had succeeded in deriving stem cells from a cloned human embryo – a process referred to as therapeutic cloning. Such cells could provide the starting point for a perfectly matched tissue graft. Or when derived from a person carrying a disease, scientists could study the cloned cells to learn more about the disease and search for drugs that modify it.

Therapeutic cloning involves reprogramming a person's DNA by injecting a skin or white blood cell nucleus into a human egg whose own DNA has been removed. But though this process was achieved in mice in 2000, and in monkeys in 2007 (by the same group), the scarcity of available human eggs had seen researchers virtually give up on humans. Especially since in 2006, the Nobel Prize-winning Japanese scientist Professor Shinya Yamanaka learned how to do it without eggs at all using iPS cells (see box: **Types of stem cells** on opposite page). A skin cell could be turned into a pluripotent stem cell merely by the insertion of four genes. Pluripotent stem cells made this way are termed induced pluripotent stem cells or iPS cells. They have revolutionised the ability to make cell models of human diseases, but researchers are more cautious about using iPS cells as tissue grafts since these cells carry a cancer-causing gene.

That's why the Oregon scientists persisted with human therapeutic cloning, building on their success with monkeys. Small tweaks to standard techniques eventually paid off. Researchers now have the option to compare the performance of iPS cells and cloned human embryonic stem cells.

ADULT STEM CELLS POWER ON

The major clinical progress with adult stem cells lies with mesenchymal stem cells (MSC) which are found in the bone marrow. These cells can regenerate connective tissue such as cartilage and bone. But they have also been found to have three other remarkable properties: they release factors that help repair diverse organs including the heart, kidney and pancreas; they tone down an overactive immune system, and if grafted into a foreign host, they do not seem to be rejected.

One of the world's most successful mesenchymal stem cell companies is Mesoblast, founded in Melbourne in 2004 by its CEO Professor Silviu Itescu. By 2011, a clinical trial suggested its stem cell product was safe and effective for treating congestive heart failure. In 45 patients who received a direct cardiac infusion of the MSCs, their hearts pumped more strongly than patients who received a standard treatment. Of the 15 who received the highest dose, none died or were hospitalised with heart failure in the following three years, while a third of those who received standard treatment did.

Mesoblast is waiting for clearance from the US authorities to start a trial later this year to conclusively establish the effectiveness of the treatment.

Trials for degenerative disc diseases are promising, too. Then there are a suite of encouraging patient trials that test the ability of MSC to temper the immune system and rescue degenerating tissue. They include early and advanced rheumatoid arthritis, type 2 diabetes and end stage diabetic kidney disease.

Meanwhile back in the lab, Australian researchers are backfilling the research pipeline with tomorrow's therapies.

It's not surprising to imagine that the breast would have a stem cell – something must be powering the massive proliferation of breast tissue during pregnancy. And in mice it's possible to take cells from breast tissue, transplant them into the fat pad of another mouse, and watch them form a breast duct. In 2006, in a world-first report published in *Nature* magazine, Professor Jane Visvader and colleagues from Melbourne's Walter and Eliza Hall Institute isolated that breast stem cell.

Visvader stresses that her work is not aimed at growing human breast tissue. Rather the focus here is treatments for breast cancer. Her research has strongly implicated the stem cell in breast cancer, particularly the hereditary form caused by inheriting a faulty gene called *BRCA1*. That finding has opened up a new treatment strategy: scientists are now looking for drugs that directly target the stem cell responsible for this cancer.

One remarkable tale of success against the odds comes from kidney researcher Professor Melissa Little at the University of Queensland. In the lab, it is one thing to cultivate embryonic stem cells to produce a single type of cell, like insulin-producing pancreatic beta cells, but the kidney is a complex filtering machine composed of many cell types. 'Everything comes from understanding how kidneys form,' Little explains. Unfortunately the mature kidney does not possess a stem cell able to make all cells of the kidney, which is part of the reason this organ is so vulnerable to failure. But embryonic kidneys certainly do.

Little's lab has been trying to reproduce this embryonic kidney stem cell starting from sources at opposite ends of the cell's life history. One starts with embryonic stem cells and painstakingly matures them towards becoming kidney cells. The other starts with adult kidney cells and shifts them back to a more immature state by treating them with 'reprograming factors'. Either way she has produced cells that resemble embryonic kidney stem cells: when she injects them into an embryonic mouse kidney – they integrate seamlessly in a way that mature kidney cells don't.

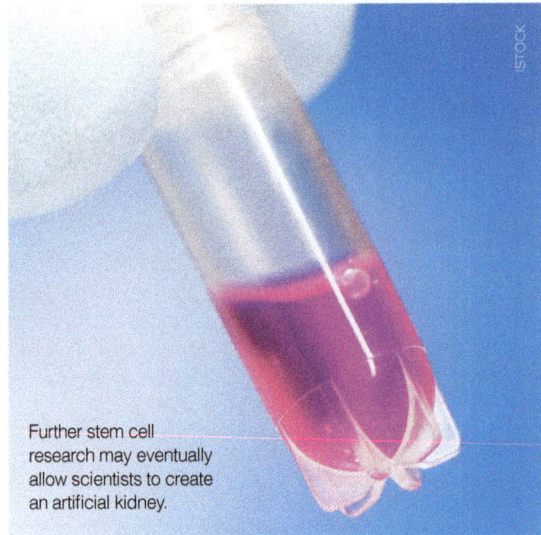

Further stem cell research may eventually allow scientists to create an artificial kidney.

Her work is about to be published in the *Journal of the American Society of Nephrology.* Little can't imagine that these cells could develop into any part of a kidney of their own accord. But they might be used to populate a kidney-shaped scaffold in a step toward building an artificial kidney. Or even sooner, they could be used to test drugs to see if they are toxic to kidney cells. "This was a risky field to go into but we are thrilled by our progress," she says.

With stem cell products in clinical trials and a research pipeline brimming with tomorrow's therapies, that's a fitting coda for 'Act One' in the great performance of stem cell research.

DR ELIZABETH FINKEL is the editor-in-chief of *Cosmos* magazine and a Vice-Chancellor's Fellow at La Trobe University.

FURTHER READING

Masahito, T. et al 2013, 'Human embryonic stem cells derived by somatic cell nuclear transfer', *Cell*, 153(6), 1228-1238.

Stem Cells Australia, 'About stem cells', http://www.stemcellsaustralia.edu.au/About-Stem-Cells.aspx.

'How to treat' pull-out supplement on stem cells, *Australian Doctor*, 18 November 2011.

Finkel, E. 2005, *Stem Cells: Controversy at the Frontiers of Science*, ABC Books, Sydney.

National Health and Medical Research Council, 'Stem cells, cloning and related issues', http://www.nhmrc.gov.au/health-ethics/human-embryos-and-cloning/stem-cells-cloning-and-related-issues.

SHUTTERSTOCK

TRUST ME – I'M A SCIENTIST

Far from being a philosophical time-waster, knowing what science is – and isn't – is vital for the health of our species, writes Michael Slezak

DANA McCAFFREY WAS 11 days old when she developed a runny nose. Less than three weeks later the little girl from northern New South Wales was dead.

"We cry ourselves to sleep with memories of our daughter coughing until she couldn't breathe," write her parents, Toni and David McCaffrey.

Dana died of whooping cough in 2009. Unable to be fully vaccinated against the disease until she was six months old, Dana's only protection was the vaccine-induced immunity of those around her. By stopping the spread of *Bordetella pertussis* – the bacterium that causes whooping cough – babies and others who cannot be vaccinated are protected by 'herd immunity'.

Nobody knows exactly where Dana contracted the bacterium, but in some parts of Australia vaccination rates are now so low – due in part to a mistaken belief that vaccination causes autism – that outbreaks of diseases such as whooping cough are a serious concern.

For a parent to decide to vaccinate their child, to urge politicians to take action on climate change, or to trust an oncologist over a miracle healer, they must be able to separate science from what's sometimes called pseudoscience. And when faced with potentially convincing claims from pseudoscience, they must be able to trust that science itself is reliable.

'Trust': it seems a funny word to associate with science, a discipline grounded in evidence and the antithesis to faith. But today we're asked to *trust* science perhaps more than any other time in history. And for good reason. The challenges of climate change, superbugs and food security demand the use of science to anticipate, manage and mitigate these risks.

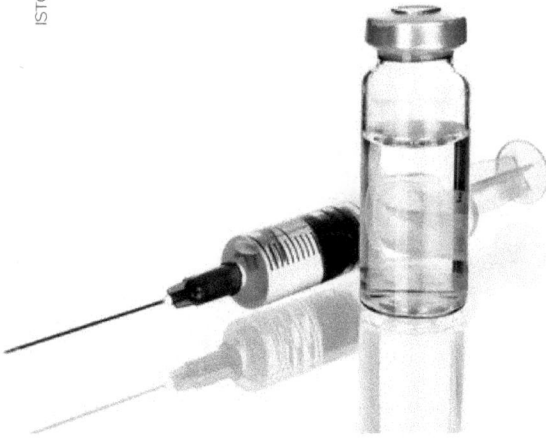

"[There are] decades of evidence that vaccination is one of the most effective developments in public health ever achieved."

We all must trust science. But this isn't the same as faith. While it's not possible to be an expert in every area, it is possible to appreciate what science is and why it's the best basis for understanding the physical and natural world.

We may not understand the evidence for a Higgs boson but we can understand why the scientific enterprise is such that we ought to believe it exists. In other words, rather than faith, we can have *good reason* to trust science.

WHAT SCIENCE IS AND ISN'T

"Kids' lives are at risk as science is ignored," read the *Sydney Morning Herald* headline. The item revealed shockingly low immunisation rates in some Sydney suburbs. But what is it, exactly, that is being ignored?

People who spread doubt about the value of vaccination often couch their doubt in the language of science. "I'm not saying that vaccines have played no role. I'm saying, let's see some evidence," demands anti-vaccination campaigner Meryl Dorey. Unfortunately, opponents like Dorey deliberately ignore decades of evidence that vaccination is one of the most effective developments in public health ever achieved.

Clearly, understanding the importance of vaccination, or any other hot-button issue, is more than just paying attention to information about diseases and immunisation. It's being able to separate *reliable* from *unreliable* information. Ask most scientists how to do this and they will reply that science deals with facts, evidence, tests and observations.

"Science is about understanding and ultimately predicting how the world works," says Professor Brian Schmidt, an Australian National University astronomer who won the

Moreover, if we trust and invest in science, the riches are likely to be enormous.

TRUST IN SCIENCE

'Trusting' science sounds, well, unscientific. At the heart of the so-called scientific method is scepticism and doubt. This supposed conflict is something with which evolutionary biologist Richard Dawkins, in his war against religious faith, is regularly challenged.

"To some, [my belief in evolution] may superficially look like faith," responds Dawkins. "But the evidence that makes me believe in evolution is not only overwhelmingly strong; it is freely available to anyone who takes the trouble to read up on it. Anyone can study the same evidence that I have and presumably come to the same conclusion."

Dawkins is a great defender of science, but the hard – and critical – question is why everyone *else* should believe in evolution, the Big Bang, anthropogenic climate change, or vaccines?

Dawkins' answer that anyone can study the same evidence and come to the same conclusion is a high bar to set. It requires knowledge and time to weigh up all the evidence for every theory. It's simply not possible for anyone – not even Dawkins – to gain that level of expertise in every scientific area.

Through observing distant supernovae, scientists discovered the accelerating expansion of the Universe. While not everyone can understand the theory behind this discovery, we have good reason to trust the science.

ISTOCK

Nobel Prize for Physics in 2011. Schmidt was awarded the prestigious prize for doing precisely this. His ground-breaking work with supernovae led to deeper understanding of dark energy: a force that permeates the cosmos. That understanding led to perhaps the ultimate of all predictions: the fate of the universe. Rather than the universe eventually collapsing in a fiery implosion, dark energy will forever push us apart, into a Big Freeze.

Many philosophers agree with Schmidt. "Science is an attempt to answer questions about the world by bringing those questions into contact with observation," explains Professor Peter Godfrey-Smith, a philosopher of science at the City University of New York Graduate Center. He adds that it's difficult to spell out exactly what that means, but it captures the basic picture of what science is. Science also moves forward. "Science is… a mixture of competition and cooperation that enables work to be cumulative. Each person takes a small step and it progresses," he says.

But when a parent is told by a chiropractor that vaccinations don't work, and that treating invisible 'subluxations' will allow the body to fight off infections successfully, how can they tell that it is bad science?

"With science you should be able to rely on the fact that there's data that is accurate and correct," says Professor Barry Marshall, a clinical microbiologist with the University of Western Australia and winner of the 2005 Nobel Prize in Physiology or Medicine.

The story of how he revolutionised the understanding of stomach ulcers illustrates precisely what some have suggested separates science from pseudoscience. It's known as 'falsificationism,' the idea that *real* science can be proved to be wrong, or 'falsified', while pseudoscience cannot.

When Marshall and his colleague Professor Robin Warren – a pathologist at Royal Perth Hospital – first argued ulcers are not caused by stress and stomach acid but by infection with the bacterium *Helicobacter pylori*, one of their harshest critics was US microbiologist Professor David Graham. "He said the great thing about the helicobacter theory of ulcers is it's going to be so easy to disprove," says Marshall. "It was a terrific concept because if it's wrong we just have to do a few experiments and we can disprove it."

Science progresses to explain newly discovered facts. The discovery that *Helicobacter pylori* infection caused stomach ulcers was revolutionary.

Of course, in the end the theory wasn't falsified and is now accepted as current scientific understanding.

Austro-British philosopher Sir Karl Popper argued in the 1950s that the ability for a theory to be falsified was the key criterion for whether something was scientific. If it wasn't possible to prove something wrong, it wasn't scientific.

But Popper was wrong. Falsificationism captured something important about science but was too blunt, sometimes painting good science as pseudoscience and vice versa. Some say falsificationism shows astrology to be scientific because it's easily falsified and indeed has been.

But worse, it seemed nothing in science could be completely falsified. In principle, any 'falsifying' evidence can be absorbed into a theory by fiddling with 'auxiliary hypotheses'. For example, when physicists in Europe incorrectly claimed to have observed faster-than-light neutrinos, many theorists proposed ways of absorbing the observation into relativity. In such cases one way to save the theory is hypothesising the measuring instruments are faulty or don't work as previously thought.

Since Popper, philosophers have proposed many ways of demarcating science from non-science: only science progresses, only scientific theories are well-tested, only science makes surprising hypotheses that turn out to be true. While each has adherents, all have failed to gain serious traction. Each appears to mismatch known examples of science and pseudoscience.

In a famous paper on the topic, philosopher Professor Larry Laudan, now at the Institute for Philosophical Research at the National Autonomous University of Mexico, concludes science is such a diverse discipline that no single criterion could possibly draw a line between science and non-science. He said we "ought to drop terms like 'pseudo-science' and 'unscientific' from our vocabulary; they are just hollow phrases which do only emotive work for us."

"Laudan gave up too quickly," counters Professor Paul Thagard, a cognitive scientist and philosopher of science at the University of Waterloo, Ontario. Rather than trying to come up with a single criterion like falsifiability – or even a set of criteria – Thagard thinks a more subtle approach is needed.

Thagard struggled to come up with strict definitions of 'science'. "What I realised later, when I got a more sophisticated view of concepts by doing cognitive science, was that concepts don't generally have strict definitions," he says,

THAGARD'S PROFILES OF SCIENCE AND PSEUDOSCIENCE

Science

- Explains using mechanisms
- Uses correlation thinking, which applies statistical methods to find patterns in nature
- Practitioners evaluate theories in relation to alternative ones
- Uses simple theories that have broad explanatory power
- Progresses over time by developing new theories that explain newly discovered facts

Pseudoscience

- Lacks mechanistic explanations
- Uses dogmatic assertions, or resemblance thinking, which infers that things are causally related merely because they are similar
- Practitioners are oblivious to alternative theories
- Uses non-simple theories that require many extra hypotheses for particular explanations
- Stagnant in doctrine and application

noting that even the simplest things don't have strict definitions. 'Restaurant', for example, could be defined as a public eating place, yet plenty of places fit that definition but aren't restaurants.

Thagard's approach – and that of many cognitive scientists – is to define concepts via 'prototypes'. A prototype has a typical set of features, none of which are essential, but together they paint a rough picture of the concept. For instance, we might not have a strict definition of 'restaurant' but our prototypical concept is more or less similar to real restaurants.

Thagard says prototypical examples of science might be parts of physics, chemistry or neuroscience. Conversely, prototypical examples of pseudoscience are astrology and creationism. Using these, Thagard developed his five features of science and pseudoscience (see: **Thagard's profiles of science and psuedoscience**).

When astrology is put through Thagard's profiles, it ticks all the pseudoscience boxes. Nobody could explain *how* the location of the stars and planets at the time of birth could affect personality or life. It's dogmatic, resistant to alternative theories and completely stagnant.

Anti-vaccination proponents might have mechanistic explanations for their beliefs, but they're very implausible. They may rely on some statistical evidence but it's cherry-picked. The proponents appear oblivious to alternative theories, dogmatically ignoring the overwhelming evidence that support them, and so on.

The debate about demarcating science from pseudoscience is far from over. Thagard's approach relies on an embryonic scientific theory of concepts and it may well need to be discarded. Regardless, Thagard argues his account is useful. These five characteristics "make a good distinction between how science works and a lot of things that pretend to be scientific," he says.

MICHAEL SLEZAK is the Australasia reporter at *New Scientist* magazine and has a background in philosophy of science.

FURTHER READING

Chalmers, A. 2013, *What Is This Thing Called Science?*, 4th edition, Open University Press, Maidenhead.

Goldacre, B. 2009, *Bad Science*, Harper Perennial, London.

Godrey-Smith, P. 2003, *Theory and Reality: An introduction to the philosophy of science*, University of Chicago Press, Chicago.

Russell, B. 1954, *The Scientific Outlook*, George Allen and Unwin, London.

Shapin, S. 1994, *A Social History of Truth*, University of Chicago Press, Chicago.

Thagard, P. 2010, *The Brain and the Meaning of Life*, Princeton University Press, Princeton.

GRAVITY WAVES MAKE YOUNG MINDS RIPPLE

Curved space and quantum weirdness make perfect sense when introduced early in school, writes **David Blair**

HERE'S MY PREDICTION: When scientists at last discover the elusive gravity waves predicted by Einstein in 1915, the event will be a landmark in the history of science. It will stand out like Heinrich Hertz's 1886 discovery of radio waves, a revelation that revolutionised the way we live and completely changed our conception of the universe.

In a nutshell, gravitational waves are ripples in the *curvature* of 'spacetime' that propagate outward from the source as a *wave*. I've spent almost 40 years trying to detect them. When I began there were just a few of us working away in university labs. Today, 1000 physicists working with billion-dollar observatories are quietly confident that the waves are within our grasp.

The problem with discussing gravity waves is they can't be described without explaining Einstein's ideas of curved space and warped time. For years I struggled to explain my work to public audiences. I wrote a popular book about gravity waves, *Ripples on a Cosmic Sea.* Then with an amazing team of volunteers and benefactors I built the $10 million Gravity Discovery Centre near Perth, filling it with exhibits designed to get Einstein's concepts across.

But despite my efforts, whenever I started to explain gravity I got glazed eyes and baffled looks.

One day I brought my son, then aged 10, to the Gravity Discovery Centre to be my assistant for a special school holiday program for top high school physics students from around Australia. The program introduced students to the concept of curved space underpinning Einstein's theory of gravity, as well as quantum weirdness which arises from Einstein's proof that light comes as photons, elementary particles of light and electromagnetic radiation. The 16 year olds were astonished by the strange ideas. But my son seemed quite relaxed. He wasn't astonished at all.

Suddenly I had a hypothesis! For those young enough to have no preformed concepts of space there are no contradictions. Sixteen year olds are exhilarated to discover a major contradiction in their education. But for adults, the contradictions are bewildering. This was the inspiration that led to the Science Education Enrichment Project.

The problem with physics education starts 2300 years ago with Euclid's book of geometry called *Elements*. This is the most influential book in the history of science, having been in print

GRAVITY WAVES HELP PROBE THE LIMITS OF SPACETIME

Einstein's theory of gravity says space is elastic and can sustain waves. Gravity waves are waves of geometry – waves in the shape of space itself – that distort the shape of all the objects they pass through. Gravity waves are unstoppable by matter. They can reveal hidden places in the universe, like the collapsing cores of exploding stars and the Big Bang itself. They can carry vast amounts of energy even when the size of the waves is minuscule.

There are thought to be 100 million black holes in our Milky Way galaxy. About 20 000 of them, along with 10 million stars, are predicted to be crammed into a tiny space only a few light years across – about the distance from here to the nearest star – around the giant black hole that lurks in the core of the Milky Way.

When two black holes get close they can capture each other, emitting vast bursts of gravitational waves as they spiral closer and closer together. Eventually, they emit a vast scream in audio frequency gravitational waves, giving out more power in gravitational wave energy than all the light power given out by all of the stars in the visible universe.

Finally, having radiated several times the mass of the Sun in pure gravitational energy, they merge into a single black hole that's born shimmering and vibrating in gravitational waves. This signature allows scientists to reconstruct the nature, mass and spin of the newly formed hole and to test Einsteinian physics in a regime where the curvature of space is like a tornado compared to the gentle breeze it is on Earth.

Gravitational waves will allow scientists to explore this hidden dark side of the universe, find out if Einstein was really right and whether a single new theory is able to unify the theory of the quantum world with the theory of space, time and gravity.

for more than 2000 years and published in more than 1000 editions. It was a basic school text for Galileo, Newton, Einstein and every educated person up to the baby boomer generation. I still have the plain slim edition that I used when I was in Year 8.

Yet today Einsteinian physics is part of our everyday reality, although most of us don't know it. Every day we use smartphones that combine exquisite quantum physics with navigation technology that corrects for time warps. With a few clicks we can send photons down optical fibres to ask Google to direct us to space telescope images of curved space in the universe, or videos showing single photons arriving one at a time to make an image.

So for people raised with Euclid, Einstein is truly weird. Some teachers claim that Einsteinian ideas are too difficult and too complex to be taught at school. Rubbish. The concepts are merely different. It is more difficult for the

Gravitational waves, predicted by Einstein, may set a 'speed limit' to rapidly spinning stars called pulsars.

teachers than for their students. Just as with spoken language it's important to start early.

Our team at the University of Western Australia and Curtin University have been conducting pilot studies with students aged 11, 12 and 16. We have created curriculum material that includes hands-on learning activities to make all the ideas vivid. Many of them are based on analogies. For instance, we use Lycra sheets for the fabric of space, and Nerf guns to create streams of photons.

Using simple graphs and Einstein's assertion that freely falling trajectories are the shortest paths in spacetime, children discover why time depends on height above the ground. They discover the quantum uncertainty principle in fun group activities. They learn to think about spacetime and appreciate that falling from a tower and floating in the space station are the same thing. They easily grasp the reality that the formulae of Euclidean geometry are approximations: really good approximations on Earth, wrong on closer inspection and completely wrong near a black hole.

The pilot studies confirm that students are neither surprised nor bewildered. The large majority don't think they are too young to learn Einsteinian ideas. We asked 16 year olds two questions: What was the most interesting concept? What was the most difficult? We found a strong correlation. The most interesting concepts were also the most difficult. This reveals that young people want to be challenged.

While today Einsteinian physics gives us our best understanding of the universe, physics still has enormous problems to solve. More than 95 per cent of the universe is mysterious dark matter and dark energy. Are black holes pathways to other universes?

A whole new spectrum of gravitational waves is waiting to be explored. Like a new great south land – we know it is there but what will we

CATCH THE WAVE

If our ears were billions of times more sensitive than they really are, we could hear gravity waves created by stellar mass black holes – black holes as heavy as stars – and neutron stars which are stars just on the point of collapse to a black hole. Space should be full of sounds like drum beats from the vibrations of black holes, rising chirps from pairs of black holes spiralling together up to hundreds of times per second, and slowly falling whistles as spinning neutron stars slow down.

The first detectors for these frequencies were huge metal bars cooled to near absolute zero (−273.15°C) which would experience the vibrations of the passing gravity waves. Five such detectors, including one in Australia, conducted searches without success. Around the year 2000, huge laser instruments that measure changes in distance between distant mirrors began to be constructed.

Gravity waves are produced at much lower frequencies if the black holes are much heavier. A tantalising prospect is to detect the gravity waves produced when galaxies merged in the early universe and the supermassive black holes in their cores themselves merged. These mergers are very slow processes because of the enormous masses involved so the frequencies are very low, like one cycle in three to 10 years. Such low-frequency signals can, in principle, be detected by replacing lasers and mirrors with radio beams arriving from very precisely spinning radio pulsars. Gravitational waves cause the distance between distant pulsars and the Earth to fluctuate so that the radio pulses come at different times.

Radio astronomers using this technique have a good chance of detecting the statistical signature of many such events, but because of the slowness of the signals it takes years to accumulate the signals and their data. Teams of astronomers across the world are pursuing this pulsar timing technique, including a very strong effort in Australia.

Between these two extremes of frequency, physicists and space scientists are developing plans for gravitational wave detectors in space, optimised for waves at a frequency of about one cycle per hour – the frequency produced when a neutron star falls into an intermediate mass black hole. Such a detector would use three spacecraft in geostationary orbits many millions of kilometres apart and laser beams to measure tiny changes in distance between them. Australian physicists are also involved in planning these missions.

discover? Who knows what will come out of it!

When Hertz was asked about the use of his discovery he said: "It's of no use whatsoever ... we just have these mysterious electromagnetic waves that we cannot see with the naked eye. But they are there."

We have thought through the implications and applications a bit better than Hertz. Yet I suspect that the best discoveries will be the surprises.

What is certain is that we will be better able to meet the challenges of the future if we allow students to begin their learning at the point of science's current best understanding, rather than condemning them to learn the old stale approximations as if they were the truth.

PROFESSOR DAVID BLAIR is the director of the Australian International Gravitational Research Centre at the University of Western Australia.

FURTHER READING

Australian International Gravitational Observatory, website, http://www.aigo.org.au/.

Blair, D. 2012, 'Testing the theory: taking Einstein to primary schools', *The Conversation*, https://theconversation.com/testing-the-theory-taking-einstein-to-primary-schools-9710.

Blair, D. and McNamara, G. 1997, *Ripples on a Cosmic Sea*, Allen & Unwin, Sydney.

The Gravity Discovery Centre, website, http://www.gravitycentre.com.au/.

The Science Show 2013, 'Eleven year olds relate to Einstein', radio broadcast, 4 May 2013, ABC Radio National.